U.S. Marine Corps Close Combat

Plus

Marine Rifle Marksmanship

Department of Defense

Pentagon
Publishing

DEPARTMENT OF THE NAVY
Headquarters United States Marine Corps
Washington, D.C. 20380-1775

18 February 1999

FOREWORD

1. PURPOSE

Today's Marines operate within a continuum of force where conflict may change from low intensity to high intensity over a matter of hours. Marines are also engaged in many military operations other than war, such as peacekeeping missions or noncombatant evacuation operations, where deadly force may not be authorized. During noncombative engagements, Marines must determine if a situation warrants applying deadly force. Sometimes Marines must decide in a matter of seconds because their lives or the lives of others depend on their actions. To make the right decision, Marines must understand both the lethal and nonlethal close combat techniques needed to handle the situation responsibly without escalating the violence unnecessarily. Marine Corps Reference Publication (MCRP) 3-02B, *Close Combat,* provides the tactics, techniques, and procedures of Marine Corps close combat. It also provides the doctrinal basis for the Marine Corps Close Combat Training Program (MCCCTP).

2. SCOPE

This publication guides individual Marines, unit leaders, and close combat instructors in the proper tactics, techniques, and procedures for close combat. MCRP 3-02B is not intended to replace supervision by appropriate unit leaders and close combat instruction by qualified instructors. Its role is to ensure standardization and execution of tactics, techniques, and procedures throughout the Marine Corps.

3. SUPERSESSION

MCRP 3-02B supersedes Fleet Marine Force Manual (FMFM) 0-7, *Close Combat,* dated 9 July 1993. There are significant differences between the two publications. MCRP 3-02B should be reviewed in its entirety.

4. WARNING

Techniques described in this publication can cause serious injury or death. Practical application in the training of these techniques will be conducted in strict accordance with approved Entry Level Close Combat, Close Combat Instructor (CCI), and Close Combat Instructor Trainer (CCIT) lesson plans. Where serious danger exists, the reac er is alerted by the following:

WARNING

5. CERTIFICATION

Reviewed and approved this date.

BY DIRECTION OF THE COMMANDANT OF THE MARINE CORPS

J. E. RHODES
Lieutenant General, U.S. Marine Corps
Commanding General
Marine Corps Combat Development Command

DISTRIBUTION: 14400006600

OVERVIEW OF CLOSE COMBAT

1. Purpose of Close Combat

Close combat is the physical confrontation between two or more opponents. It involves armed and unarmed and lethal and nonlethal fighting techniques that range from enforced compliance to deadly force. The purpose of close combat is to execute armed and unarmed techniques to produce both lethal and nonlethal results. Unarmed techniques include hand-to-hand combat and defense against hand-held weapons. Armed techniques include techniques applied with a rifle, bayonet, knife, baton, or any weapon of opportunity.

2. Continuum of Force

Marines will find themselves in both combative and noncombative situations. The threat level in these situations can rise and fall several times based on the actions of both Marines and the people involved. The escalation of force stops when one person complies with the demands imposed by another person. This range of actions is known as a continuum of force. Continuum of force is the concept that there is a wide range of possible actions, ranging from voice commands to application of deadly force, that may be used to gain and maintain control of a potentially dangerous situation (MCO 5500.6_, *Arming of Security and Law Enforcement [EE] Personnel and the Use of Force).* Continuum of force consists of five levels that correspond to the behavior of the people involved and the actions Marines use to handle the situation (see the table below). Close combat techniques are executed in levels three, four, and five.

Level One: Compliant(Cooperative)

The subject complies with verbal commands. Close combat techniques do not apply.

Level Two: Resistant (Passive)

The subject resists verbal commands but complies immediately to any contact controls. Close combat techniques do not apply.

Continuum of Force		
Level	**Description**	**Actions**
1	Compliant (Cooperative)	Verbal commands
2	Resistant (Passive)	Contact controls
3	Resistant (Active)	Compliance techniques
4	Assaultive (Bodily Harm)	Defensive tactics
5	Assaultive (Serious Bodily Harm/ Death)	Deadly force
Note: Shading indicates levels in which Marines use close combat techniques.		

Level Three: Resistant (Active)

The subject initially demonstrates physical resistance. Marines use compliance techniques to control the situation. Level three incorporates close combat techniques to physically force a subject to comply. Techniques include—

- Come-along holds.
- Soft-handed stunning blows.
- Pain compliance through joint manipulation and the use of pressure points.

Level Four: Assaultive (Bodily Harm)

The subject may physically attack Marines, but he does not use a weapon. Marines use defensive tactics to neutralize the threat. Defensive tactics include the following close combat techniques:

- Blocks.
- Strikes.
- Kicks.
- Enhanced pain compliance procedures.
- Nightstick blocks and blows.

Level Five: Assaultive (Serious Bodily Harm/Death)

The subject usually has a weapon and will either kill or seriously injure someone if he is not stopped immediately and brought under control. Typically, to control the subject, Marines apply deadly force through the use of a firearm, but they may also use armed and unarmed close combat techniques.

3. Marine Corps Tactical Concepts

Close combat techniques support the following key Marine Corps tactical concepts. The concepts are not standalone ideas but are to be combined to achieve an effect that is greater than their separate sum.

Achieving a Decision

Achieving a decision is important in close combat. An indecisive fight wastes energy and possibly Marines' lives. Whether the intent is to control an opponent through restraint or defend themselves in war, Marines must have a clear purpose before engaging in close combat and act decisively once engaged.

Gaining an Advantage

A basic principle of martial arts is to use the opponent's strength and momentum against him to gain more leverage than one's own muscles alone can generate, thereby gaining an advantage. In close combat, Marines must exploit every advantage over an opponent to ensure a successful outcome. This can include employing various weapons and close combat techniques that will present a dilemma to an opponent. Achieving surprise can also greatly increase leverage. Marines try to achieve surprise through deception, stealth, and ambiguity.

Speed

Marines use speed to gain the initiative and advantage over the enemy. In close combat, the speed and violence of the attack against an opponent provides Marines with a distinct advantage. Marines must know and understand the basics of close combat so they can act instinctively with speed to execute close combat techniques.

Adapting

Close combat can be characterized by friction, uncertainty, disorder, and rapid change. Each situation is a unique combination of shifting factors that cannot be controlled with precision or certainty. For example, a crowd control mission may call for Marines to employ various techniques ranging from nonlethal restraint to more forceful applications. Marines who adapt quickly will have a significant advantage.

Exploiting Success

Typically, an enemy will not normally surrender simply because he was placed at a disadvantage. Marines cannot be satisfied with gaining an advantage in a close combat situation. They must exploit any advantage aggressively and ruthlessly until an opportunity arises to completely dominate the opponent. Marines must exploit success by using every advantage that can be gained.

(reverse blank)

Close Combat

Table of Contents

CHAPTER 1

FUNDAMENTALS OF CLOSE COMBAT

This chapter describes all techniques for a right-handed person. However, all techniques can be executed from eittwr side.

The Marine is depicted in camouflage utilities. The opponent is depicted without camouflage.

The fundamentals of close combat include ranges, weapons of the body, target areas of the body, and pressure points of the body. These fundamentals form the basis for all close combat techniques. They provide Marines with a common framework regardless of the type of confrontation or the techniques used. If Marines apply these fundamentals properly in a close combat situation, they may save their lives or the lives of fellow Marines.

1. Ranges of Close Combat

Close combat engagements occur within three ranges: long range, midrange, and close range. During any engagement, these ranges may blur together or may rapidly transition from one to another until either the opponent is defeated or the conflict is resolved.

Long Range

During long range engagements, combatants engage each other with rifles, bayonets, sticks, or entrenching tools. See figure below.

Midrange

During midrange engagements, combatants engage each other with knives, punches, or kicks.

Close Range

During close range engagements, combatants grab each other. Close range engagements also involve elbow strikes, knee strikes, and grappling.

2. Weapons of the Body

Hands and Arms

The hands, forearms, and elbows are the arm's individual weapons. The hands consist of several areas that can be used as weapons: fists, edges of hands, palms, and fingers.

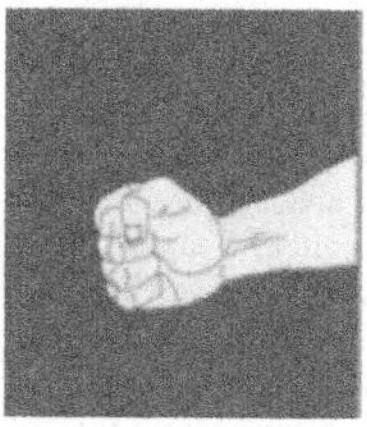

Fists. To minimize injury to the fists, Marines use their fists as weapons to target soft tissue areas such as the throat. The fists' striking surfaces are the first two knuckles of the hands or the meaty portions of the hands below the little fingers.

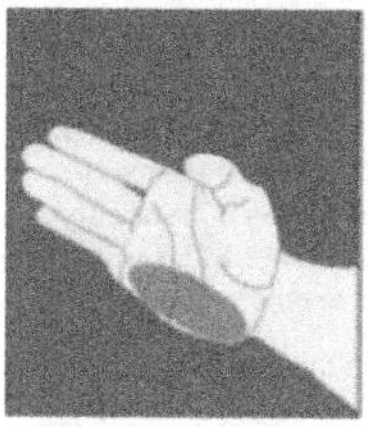

Edge of Hand. Marines use the edge of the hand (knife edge) as a weapon. Marines use the edge of the hand to strike soft tissue areas.

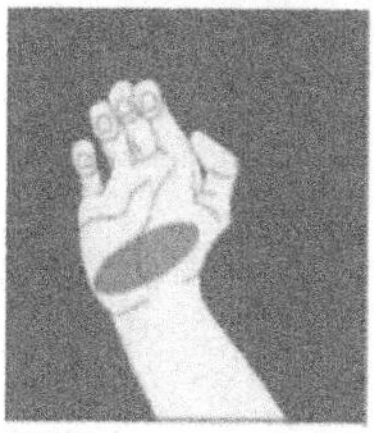

Palms. Because of the palm's padding, Marines use the heels of the palms to strike, parry, and/or block.

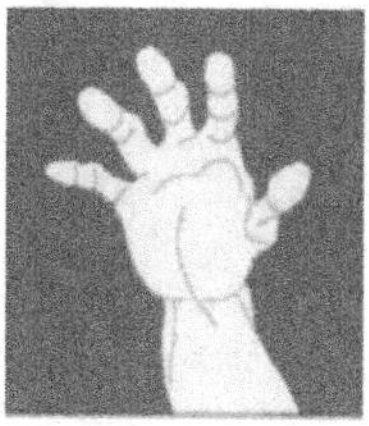

Fingers. Marines use the fingers to gouge, rip, and tear soft tissue areas (e.g., eyes, throat, groin).

Forearms. Marines use the forearms as a defensive tool to deflect or block attacks. Forearms can also be used as striking weapons to damage or break an opponent's joints and limbs. Marines sustain less self-injury when strikes are conducted with the forearms than when strikes are conducted with fists and fingers.

Elbows. Marines use the elbows as striking weapons. Because of the short distance needed to generate power, elbows are excellent weapons for striking during the close range of close combat.

Legs

The legs are more powerful than any other weapon of the body, and they are less prone to injury when striking. The feet are protected by boots and are the preferred choice for striking.

Feet. Marines use the balls of the feet, the insteps, and the toes to kick an opponent. Marines use the cutting edge of the heels and the heels to stomp on an opponent. Marines must be wearing boots when striking with the toes.

Knees. Like elbows, knees are excellent weapons in the close range of close combat. Knee strikes are most effective while fighting close to an opponent where kicks arc impractical. The opponent's groin area is an ideal target for the knee strike if he is standing upright. Knee strikes can deliver a devastating secondary attack to an opponent's face following an initial attack that caused him to bend at the waist.

3. Target Areas of the Body

During close combat, Marines strive to attack the accessible target areas of an opponent's body. The readily accessible areas will vary with each situation and throughout the engagement. The target areas are divided into five major groups: head, neck, torso, groin, and extremities. The figure below illustrates target areas of the body.

Head

The vulnerable regions of the head are the eyes, temple, nose, ears, and jaw. Massive damage to the head kills an opponent.

Eyes. The eyes are excellent targets because they are soft tissue areas that arc not protected by bone or muscle. Attacks to this area may cause the opponent to protect the area with his hands, allowing Marines to execute a secondary attack to other

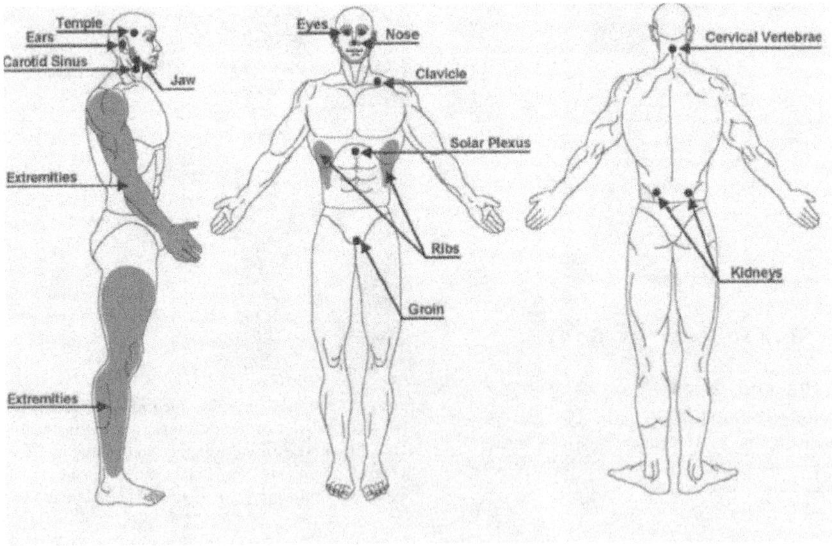

target areas while the opponent uses his hands to protect his eyes.

Temple. The temple is one of the most fragile areas of the head. Powerful strikes to the opponent's temple cause permanent damage and death.

Nose. The nose is very sensitive and easily broken. An attack to this area causes involuntary watering and closing of the opponent's eyes, rendering him vulnerable to secondary attacks. However, through training, individuals can condition themselves to withstand attacks to the nose. Therefore, any attack to the nose must be powerfully delivered.

Ears. Attacks to the ears may cause the eardrum to rupture. But this may not stop or even distract an opponent unless Marines powerfully deliver the strike.

Jaw. The jaw region, when struck forcefully, renders the opponent unconscious. Strikes to the jaw cause painful injuries to the teeth and surrounding tissues (e.g., lips, tongue), but the risk of self-injury is great unless Marines deliver strikes with a hard object such as a helmet, rifle butt, or boot heel.

Neck

The front of the neck, or throat area, is a soft tissue area that is not covered by natural protection. Damage to this region causes the opponent's trachea to swell, closing his airway, which can lead to death.

Carotid Sinus. The carotid sinus is located on both sides of the neck just below the jaw. Strikes to the carotid sinus restrict blood flow to the brain, causing loss of consciousness or death.

Cervical Vertebrae. The cervical vertebrae on the back of the neck, from the base of the skull to the top of the shoulders, contains the spinal cord, which is the nervous system's link to the brain. The weight of the head and the lack of large muscle mass allow damage to the cervical vertebrae and spinal cord. Excessive damage to this area causes pain, paralysis, or death.

Torso

Clavicle. The opponent's clavicle (or collar bone) can be easily fractured, causing immobilization of the arm.

Solar Plexus. Attacks to the opponent's solar plexus or center of the chest can knock the breath out of him and immobilize him.

Ribs. Damage to the opponent's ribs immobilizes him. It may also cause internal trauma.

Kidneys. Powerful attacks to the opponent's kidneys cause immobilization, permanent damage, or death.

Groin

The groin area is another soft tissue area not covered by natural protection. Any damage to this area causes the opponent to involuntarily protect his injured area, usually with his hands or legs. In male opponents, the scrotum is the main target since even a near miss causes severe pain, contraction of the lower abdominal muscles, deterioration of his stance, and possible internal trauma.

Extremities

Rarely will an attack to the opponent's extremities (arms and legs) cause death, but they are important target areas in close combat. Damage to an opponent's joints causes immobilization.

4. Pressure Points of the Body

There are nerves in the human body that, when pressure is applied or when they are struck, allow Marines to control a subject through pain compliance. Marines use pressure points to control an opponent when deadly force is not authorized. They also use pressure points to soften or distract an opponent so a lethal or nonlethal technique can be employed. The figure on page 1-5 illustrates the body's pressure points. Marines execute attacks to pressure points by—

: Rapidly kicking or striking pressure points.
: Slowly applying steady pressure to pressure points.

Infraorbital Nerve

The infraorbital nerve is just below the nose. Marines apply pressure to this nerve with an index finger to control the opponent.

Mastoid Process

The mastoid process is behind the base of the ear and beneath the edge of the jaw. Marines apply inward and upward pressure to this pressure point with the fingers to distract and control the opponent.

Jugular Notch

The jugular notch is at the base of the neck in the notch formed at the center of the clavicle. Marines apply pressure in a quick, stabbing motion with the index finger. Strikes to the jugular notch cause serious damage.

Brachial Plexus (Tie In)

The brachial plexus (tie in) is on the front of the shoulder at the joint. Strikes and pressure applied with the hand are effective on this nerve.

Radial Nerves

Radial nerves are on the inside of the forearms along the radius bones. Strikes and pressure applied with the hand to the radial nerve serve as a softening technique.

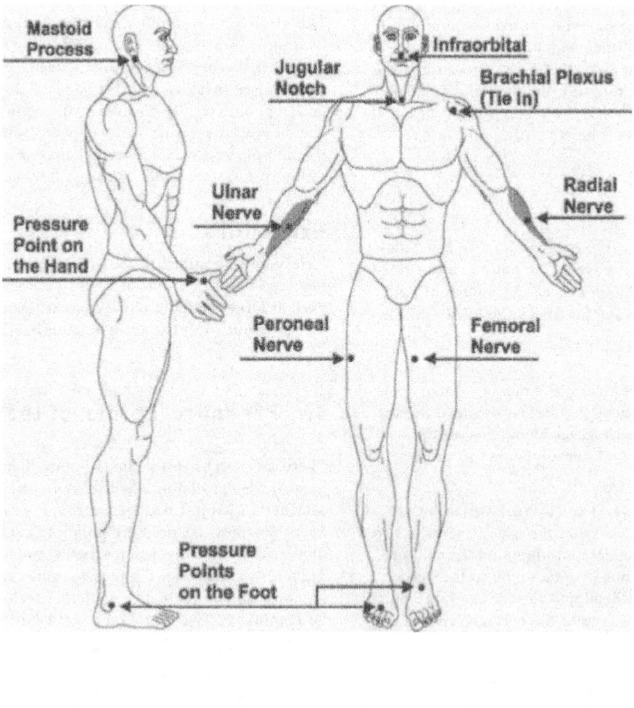

Ulnar Nerve

Ulnar nerves are on the outside of the forearms along the ulnar bones. Strikes and pressure applied with the hand to the ulnar nerve serve as a softening technique.

Pressure Point on the Hand

The hands contain a pressure point on the webbing between the thumbs and index fingers where the two bones of the fingers meet.

To force an opponent to soften or release his grip, Marines apply pressure with their index fingers to this pressure point or strike this pressure point with their fists.

Femoral Nerves

Femoral nerves are on the inside of the thighs along the femur bones. Strikes to the femoral nerve serve as a softening technique.

Peroneal Nerves

Peroneal nerves are on the outside of the thighs along the femur bones. Strikes to the peroneal nerve serve as a softening technique.

Pressure Points on the Feet

There are pressure points on the feet that, when pressure is applied or when they are struck, serve to soften or distract the opponent. Marines apply pressure with the toe, edge, or heel of their boots to the following points:

: The notch below the ball of the ankle.
: The top center of the foot, above the toes.
: The top of the foot where the leg and foot meet.

5. Basic Warrior Stance

Achieving a solid stance is critical to stability and movement throughout any close combat confrontation. The basic warrior stance provides the foundation for stability and movement that is needed to execute close combat techniques. To execute the basic warrior stance, Marines put their feet apart, hands up, elbows in, and chin down.

Feet Apart

Place feet shoulder-width apart.

Keep the head forward and eyes on the opponent, take a half step forward with the left foot, and pivot on the heels so the hips and shoulders are at approximately a 45-degree angle to the right.

Distribute body weight evenly on both legs. Bend the knees slightly.

Hands Up

Curl the fingers naturally into the palm of the hand. Position the thumb across the index and middle fingers. Do not clench the fists. Clenching the fists increases muscular tension in the forearms and decreases speed and reaction time.

Bring the hands up to the face at chin level, with the palms facing each other. Hold the fists up high enough to protect the head, but not so high that they block the field of vision. Ensure continuous eye contact with the opponent.

Elbows In

Tuck the elbows in close to protect the body.

Chin Down

Tuck the chin down to take advantage of the natural protection provided by the shoulders.

6. Angles of Approach and Movement

Marines use movement to control a confrontation and to retain a tactical advantage. Movement increases power and maximizes momentum. By moving around the opponent, Marines gain access to different target areas of the opponent's body. Movement allows Marines to use different weapons of the body and different close combat techniques to attack specific target areas.

Angles of Approach

Marines move anywhere within a 360-degree circle around the opponent to gain a tactical advantage. This circle provides access to different target areas of the opponent's body.

When facing an opponent, Marines move in a 45-degree angle to either side of the opponent. Moving at a 45-degree angle avoids an opponent's strike and puts Marines in the best position to attack the opponent. Marines should avoid being directly in front of an opponent during a confrontation. If a Marine is directly in front of an opponent, the opponent can rely on his forward momentum and linear power to seize the tactical advantage.

Movement

Marines must know how to move in all directions while maintaining the basic warrior stance. During any movement, the legs or feet should not be crossed. Once a movement is completed, the basic warrior stance should be resumed. Maintaining the basic warrior stance protects Marines and puts them in the proper position to launch an attack against an opponent.

Note: Before body movement begins. Marines turn their heads quickly to the new direction. The faster the head turns, the faster the body moves, and the quicker Marines attain visual contact with the opponent.

Forward to the Left. To move forward to the left from the basic warrior stance, Marines-

: Move the left foot forward at a 45-degree angle from the body (approximately 12 to 15 inches), keeping the toe pointed toward the opponent.

Bring the right foot behind the left foot as soon as the left foot is in place. This returns Marines to the basic warrior stance.

Forward to the Right. To move forward to the right from the basic warrior stance, Marines-

Move the right foot forward at a 45-degree angle from the body (approximately 12 to 15 inches).

Bring the left foot, toe pointing toward the opponent, in front of the right foot as soon as the right foot is in place. This returns Marines to the basic warrior stance.

Backward to the Left. To move backward to the left from the basic warrior stance, Marines execute the forward movement in reverse. Marines-

Move the left foot backward at a 45-degree angle from the body (approximately 12 to 15 inches), keeping the toe pointed toward the opponent.

Bring the right foot behind the left foot as soon as the left foot is in place. This returns Marines to the basic warrior stance.

Backward to the Right. To move backward to the right from the basic warrior stance, Marines execute the forward movement in reverse. Marines-

Move the right foot backward at a 45-degree angle from the body (approximately 12 to 15 inches).

Bring the left foot, toe pointing toward the opponent, in front of the right foot as soon as the right foot is in place. This returns Marines to the basic warrior stance.

7. Balance and Off-Balancing

Balance

In any close combat situation, Marines must strive to maintain balance. The last place to be in a close combat situation is on the ground. Marines must maintain a strong base and a low center of balance, their feet must be a shoulder-width apart, and they must stay on their toes to enable quick movement.

Off-balancing

Marines use off-balancing techniques to control an opponent. These techniques are used to throw an opponent to the ground while Marines remain standing, or they are used to put Marines in a position for an offensive attack.

Off-balancing techniques use the opponent's momentum to move or throw him. For example, if the opponent is charging a Marine, the Marine pulls the opponent to drive him to the ground. Likewise, if the opponent is pulling a Marine, the Marine pushes the opponent to drive him to the ground.

Off-balancing techniques also rely on the power generated by the opponent. For example, during combat a Marine may be tired or outnumbered. Depending on the generated energy and momentum of the opponent, the Marine employs off-balancing techniques with very little effort and still provides effective results.

Because off-balancing techniques rely on the momentum and power generated by the opponent, these techniques are particularly effective for Marines who may be outsized by their opponent or lack their opponent's strength.

Angles of Off-balancing

There are eight angles or directions in which an opponent can be off-balanced: forward, rear, right, left, forward right, forward left, rear right, and rear left.

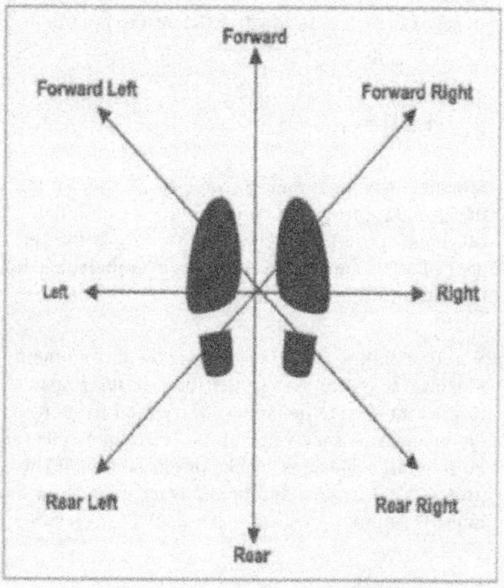

Note: The angles correspond to the Marine's perspective, not the opponent's. Forward, rear, right, and left are straight angles. Forward right, forward left, rear right, and rear left are considered quadrants that are at a 45-degree angle in either direction to the front or rear.

Off-balancing Techniques

Marines off-balance an opponent by pushing, pulling, or bumping the opponent with their hands, arms, or bodies.

To pull, Marines grab an opponent with their hands and drive him forcefully to one of the rear quadrants or to the right or left.

To push, Marines grab the opponent with their hands and drive him forcefully into one of the front quadrants or to the right or left.

Marines execute bumping in the same manner as pushing, but use their shoulders, hips, and legs instead of their hands to off-balance the opponent.

8. Falls

Marines may lose their balance or be thrown to the ground during encounters with an opponent. Marines use falling techniques to absorb the impact of a fall and to quickly return to their feet following an opponent's attack.

Whether falling or being thrown by an opponent, Marines strive to reduce the force of the impact, to prevent serious personal injury, and to increase the chances of survival. Falling techniques use the body's large muscles (back, thighs, buttocks) to protect vital organs and bones from injury and immobilization.

Front Fall

Marines execute a front fall to break a fall on the front. To execute the front fall, Marines-

Bend the elbows and place the palms facing out in a position to spread and absorb the impact of the fall.

Fall forward, breaking the fall with the forearms and palms. The forearms and hands, down to the fingertips, should strike the ground simultaneous-iy-

Offer resistance with the forearms and hands to keep the head raised off the ground.

Side Fall

Marines execute a side fall to break a fall on the side. To execute the side fall, Marines-

Bring the right arm across the body so the hand is next to the left shoulder with the palm facing inboard.

Fall to the side, breaking the fall with the right arm by slapping the ground and making contact from the shoulder or forearm down to the hand. At the same time, tuck the chin and keep the head raised off the ground. The chin should be tucked to the chest at all times to prevent whiplash.

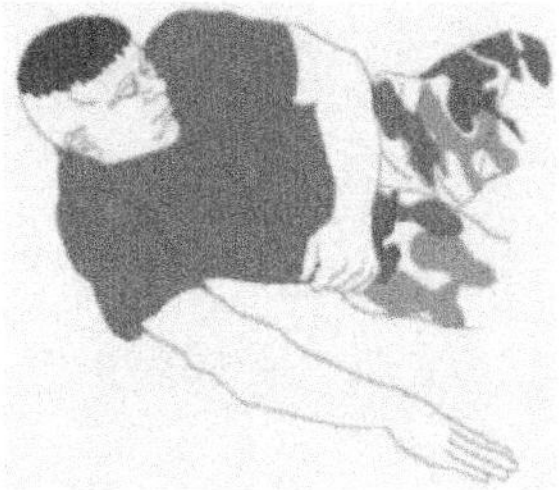

Stretch the right leg out to make contact with the ground and to distribute and absorb the impact. Bend the left leg, allowing the foot to make contact with the ground.

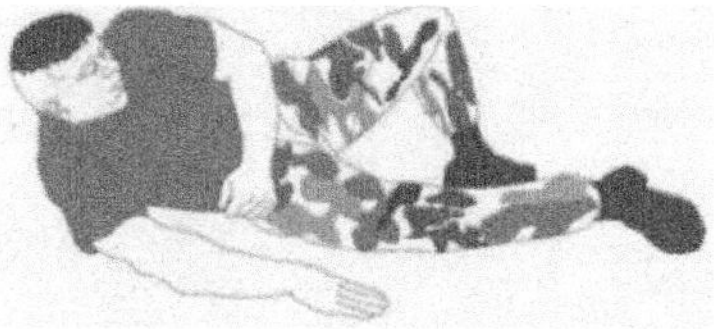

Back Fall

Marines execute a back fall to break the fall when being thrown or falling backward. To execute the backfall, Marines-

Cross the hands in front of the chest and tuck the chin.

Fall backward and slap the ground with the forearms and hands to absorb the impact of the fall and keep the head off the ground.

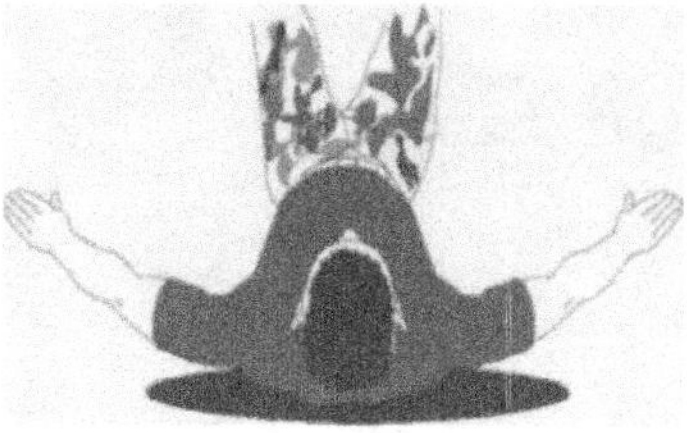

Forward Shoulder Roll

Marines use the forward shoulder roll to break a fall from an opponent's attack and to use the momentum of the fall to get back on their feet quickly. Ideally, Marines execute the forward shoulder roll to a standing position so they can continue fighting. To execute the forward shoulder roll, Marines-

Contact the ground with the back of the right forearm and upper arm. Tuck the chin into the chest.

Roll onto the right shoulder, rolling diagonally across the back to land on the left hip.

Slap the ground with the left arm, absorbing the impact from the shoulder to the hand, palm down.

Keep the left leg straight to absorb as much of the impact as possible. The right leg is bent and the foot hits flat on the ground.

Bend the left leg upon impact to push off with the left knee and leg to a squatting and then a standing position. Forward momentum should carry the Marine to a standing position.

CHAPTER 2

LETHAL AND NONLETHAL WEAPONS TECHNIQUES

This chapter describes all techniques for a right-handed person. However, all techniques can be executed from either side.

In drawings, the Marine is depicted in woodland camouflage utilities; the opponent is depicted without camouflage. In photographs, the Marine is depicted in woodland camouflage utilities; the opponent is depicted in desert camouflage utilities.

1. Bayonet Techniques

WARNING

During training, Marines must have bayonets sheathed. Marines use bayonet dummies to practice bayonet techniques. When practicing offensive and defensive bayonet techniques student-on-student, Marines use pugil sticks.

All Marines armed with a rifle carry a bayonet. The bayonet is an effective weapon if Marines are properly trained in offensive and defensive bayonet techniques. An offensive attack, such as a thrust, is a devastating attack that can quickly end a fight. Defensive techniques, such as the block and parry, can deter the opponent's attack and allow Marines to regain the initiative. Through proper training, Marines develop the courage and confidence required to effectively use a bayonet to protect themselves and destroy the enemy. In situations where friendly and enemy troops are closely mingled and rifle fire and grenades are impractical, the bayonet becomes the weapon of choice.

Holding the Rifle

To execute bayonet techniques, Marines hold the rifle in a modified basic warrior stance. All movement begins and ends with the basic warrior stance. To hold the rifle, Marines—

Use an overhanded grasp to grab the small of the rifle's stock. Use an underhanded grasp to grab the handguards of the rifle.

Lock the buttstock of the rifle against the hip with the right forearm.

Orient the blade end of the rifle toward the opponent.

Offensive Bayonet Techniques

Straight Thrust. Marines use the straight thrust to disable or kill an opponent. It is the most deadly offensive technique because it causes the most trauma to an opponent. Target areas are the opponent' s throat, groin, or face. The opponent's chest and stomach are also excellent target areas if not protected by body armor or combat equipment. To execute the straight thrust, Marines—

Lift the left leg and lunge forward off the ball of the right foot while thrusting the blade end of the weapon forward, directly toward the opponent.

Retract the weapon and return to the basic warrior stance.

Slash. Marines use the slash to kill an opponent or to create an opening in his defense. The target area is the opponent's neck. To execute the slash, Marines—

Extend the left hand back toward the left shoulder.

Thrust the left hand forward and swing it to the right, bring the right hand back toward the hip, and turn the cutting edge of the blade toward the opponent's neck. The movement is a slashing motion so the blade cuts across the opponent's neck.

Horizontal Buttstroke. Marines use the horizontal buttstroke to weaken an opponent's defenses, to cause serious injury, or to set him up for a killing blow. Target areas are the opponent's head, neck, and legs. To execute the horizontal buttstroke, Marines—

Step forward with the right foot and drive the right hand forward. Rotate the hips and shoulders into the strike. Move the left hand back toward the left shoulder.

Strike the opponent with the butt of the weapon.

Vertical Buttstroke. Marines use the vertical buttstroke to weaken an opponent's defenses, to cause serious injury, or to set him up for a killing blow. Target areas are the opponent's groin and face. To execute the vertical buttstroke, Marines—

Step forward with the right foot and drive the right hand straight up.

Pull the left hand back over the left shoulder.

Strike the opponent with the butt of the weapon.

Smash. Marines use the smash as a follow-up technique to the horizontal or vertical buttstroke, particularly if they missed the target. The target area is the opponent's head. To execute the smash following a buttstroke, Marines-

Step forward with the right foot and place the blade end of the weapon over the left shoulder and elevate the right elbow above the shoulders.

Drive the arms straight forward, striking the opponent with the butt of the weapon.

Defensive Bayonet Techniques

Parry. Marines use a parry as a defensive technique to redirect or deflect an attack. A parry is a slight redirection of a linear attack by an opponent; e.g., a straight thrust or a smash. To execute the parry, Marines-

Use the bayonet end of the rifle to redirect the barrel or bayonet of the opponent's weapon.

Lock the weapon against the hip with the right forearm.

Rotate to the right or left, moving the bayonet end of the rifle to parry the opponent's attack. Rotation is generated from the hips.

Redirect or guide the opponent's weapon away from the body by exerting pressure against the opponent' s weapon.

High Block. Marines execute a high block against a vertical attack coming from high to low. To execute the high block, Marines—

Thrust the arms up forcefully at approximately a 45-degree angle from the body. The weapon should be over the top of the head and parallel to the ground. The elbows are bent, but there is enough muscular tension in the arms to absorb the impact and deter the attack.

Low Block. Marines execute the low block

against a vertical attack coming from low to high. To execute the low block, Marines—

Thrust the arms down forcefully at approximately a 45-degree angle from the body. The weapon

should be at or below the waist and parallel to the ground. The elbows are bent, but there is enough muscular tension in the arms to absorb the impact and deter the attack.

Left and Right Block. Marines execute a left or right block against a horizontal buttstroke or a slash. To execute the left or right block, Marines-

Thrust the arms forcefully to the right or left, holding the rifle vertically in the direction of the attack. The elbows are bent, but there is enough muscular tension in the arms to absorb the impact and deter the attack.

Counter Action Following the Block. After deflecting an opponent's attack with a block, Marines counter with a slash or a horizontal buttstroke to regain the initiative. However, the objective in any bayonet fight is to thrust forward with the blade end of the weapon to immediately end the fight.

Group Strategy

On occasion, Marines may engage an opponent as a member of a group or numerous opponents by one's self or as a member of a group. By combining bayonet fighting movements and simple strategies, Marines can effectively overcome their opponent or opponents.

Offensive Strategy: Two Against One. If two bayonet fighters engage one opponent, the fighters advance together.

Fighter 1 engages the opponent while fighter 2 swiftly and aggressively attacks the opponent's exposed flank and destroys the opponent.

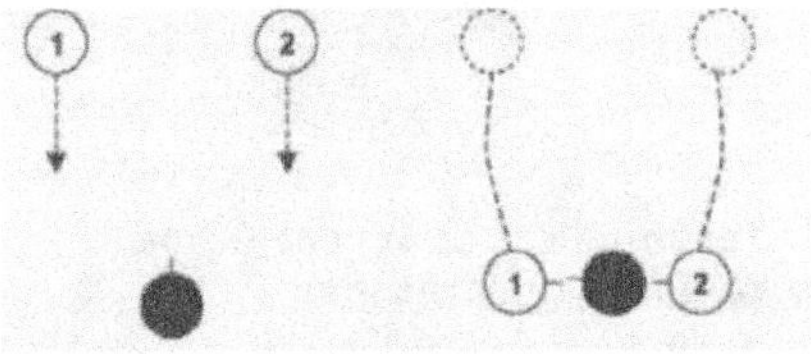

Offensive Strategy: Three Against Two. If three bayonet fighters engage two opponents, the fighters advance together keeping their opponents to the inside.

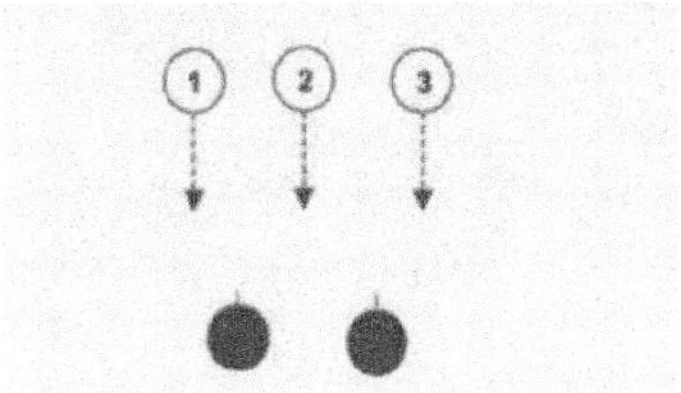

Fighters 1 and 3 engage opponents. Fighter 2 attacks the opponent's exposed flank engaged by fighter 1 and destroys the opponent.

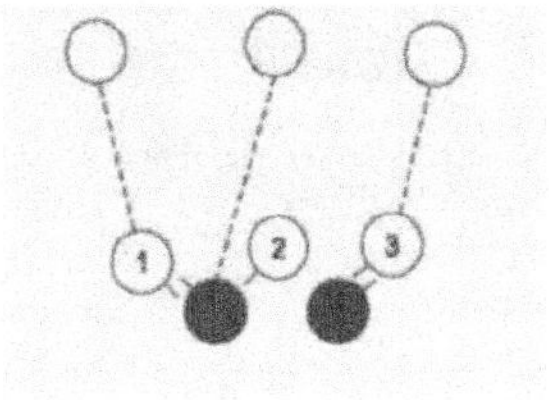

Fighters 1 and 2 turn and attack the exposed flank of the opponent engaged by fighter 3 and destroy the opponent.

Defensive Strategy: One Against Two. If a fighter is attacked by two opponents, the fighter immediately positions himself at the flank of the nearest opponent and keeps that opponent between himself and the other opponent.

Using the first opponent's body as a shield against the second opponent, the fighter destroys the first opponent quickly before the second opponent moves to assist.

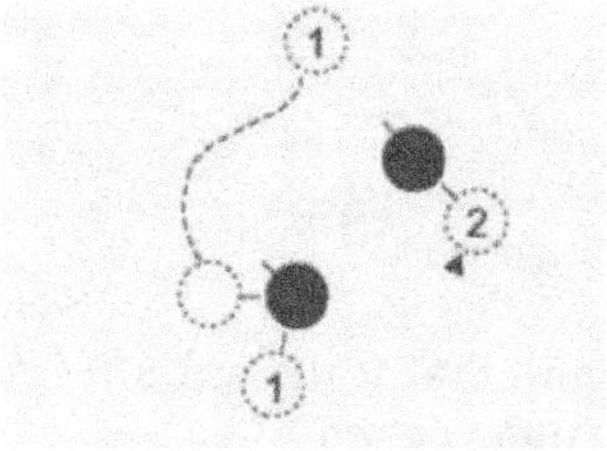

Then, the fighter engages and destroys the second opponent.

Defensive Strategy: Two Against Three. If two fighters are attacked by three opponents, the fighters immediately move to the opponent's flanks.

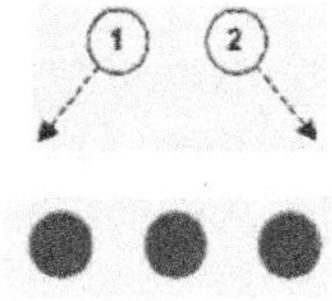

Fighters 1 and 2 quickly attack and destroy their opponents before the third opponent closes in.

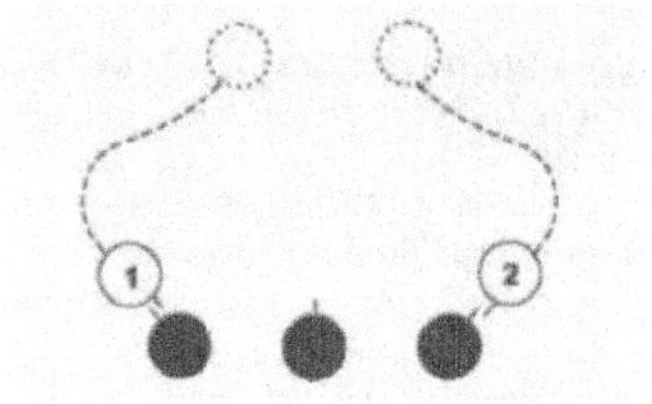

Fighter 1 engages the third opponent while fighter 2 attacks the opponent's exposed flank and destroys the opponent.

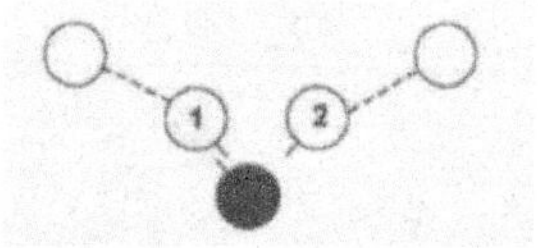

2. Nonlethal Rifle and Shotgun Retention Techniques

Most Marines are armed with the M16A2 service rifle. Marines are taught to keep their weapon with them at all times. Marines must be constantly alert to their surroundings and the people moving in and around their environment. Marines may be confronted by an individual who tries to take theii weapons. If this happens, Marines should not struggle with the individual. To retain positive control of their weapons, Marines must understand and apply weapons retention techniques, otherwise known as armed manipulation. The fol lowing techniques can be used with either the rifle or the shotgun.

Blocking Technique

To execute a blocking technique, Marines-

Stand in a defensive posture.

Use the weapon to block the opponent by thrusting it out firmly, with the elbows bent. Do not try to hit the opponent with the rifle. The rifle is used as a barrier.

Technique if Opponent Grabs Weapon Underhanded

Usually an opponent will try to grab the weapon or block it as an instinctive action. If the opponent uses an underhand grab to seize the handguards of the weapon, Marines—

Trap the opponent's closest finger(s) above the knuckle with the thumb so he cannot release his grip-

Apply bone pressure on the opponent's finger to initiate pain compliance.

Rotate the barrel of the weapon up or down quickly while maintaining pressure on the opponent's hand. At the same time, quickly pivot to off-balance the opponent.

Technique if Opponent Grabs Weapon Overhanded

If the opponent uses an overhand grab to seize the weapon overhanded, Marines—

Trap the opponent's finger to hold his hand in place.

Rotate the barrel to place it across the opponent's forearm and apply downward pressure. This action is similar to an armbar.

Technique if the Opponent Grabs the Muzzle

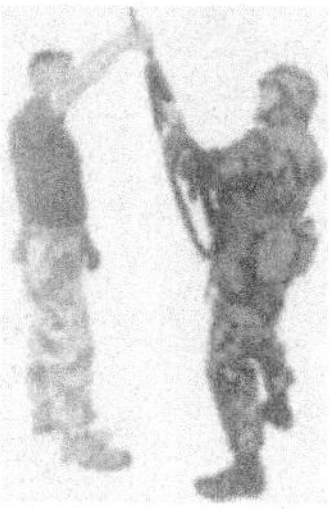

If the opponent grabs the muzzle of the weapon, Marines—

Rotate the muzzle quickly in a circle motion, Slash downward with the muzzle to release the opponent's grip.

Butt Strikes

Strikes with the butt of the weapon control or ward off an attacker. During any of the retention techniques, Marines use the heel or cutting edge of the weapon to deliver butt strikes to the inside or the outside of the opponent's thigh. The inside butt strike targets the femoral nerve as illustrated in the photo to the left.

The outside butt strike targets the peroneal nerve as illustrated in the photo to the left. Strikes can be made to the outside or inside of the thighs. If a strike to one side of the thigh misses, Marines follow back through with the butt of the weapon on the other side of the thigh.

Off-balancing Techniques

Marines apply off-balancing techniques to throw an opponent to the ground and retain possession of the weapon.

If the opponent grabs the weapon and pushes, Marines should not push on the weapon. They should—

Move with the momentum and movement of the

opponent by pivoting in the direction of the movement by stepping back.

Throw the opponent to the ground with a quick jerking movement by lowering the mu/./le and swinging the butt of the weapon.

If the opponent grabs the weapon and pulls, Marines—

Step on the opponent's foot and push forward to off-balance him and drive him to the ground.

Sweep the opponent's feet out from under him by hooking his leg with the leg and kicking backward.

3. Nonlethal Handgun Retention Techniques

Many Marines are armed with the M9 service pistol. Marines must keep their weapons in their possession at all times. Marines must be constantly alert to their surroundings and the people moving in and around the environment. Marines may be confronted by an individual who tries to take their weapons. To retain positive control of the weapon, Marines must understand and apply handgun retention techniques.

Blocking Technique

Marines perform the following blocking techniques if an opponent attempts to grab their pistol in the holster. Marines—

Place the body between the weapon and the opponent by immediately pivoting so the weapon is away from the opponent.

Step back and away from the opponent while placing the hand on the pistol grip.

Extend the left hand and block, deflect, or strike the opponent's arm.

Armbar Technique

Marines use the armbar technique when an opponent uses his right hand to grab their pistol in the holster. To execute the armbar technique, Marines—

Trap the opponent's right hand by grasping the opponent's wrist or hand with the right hand and applying pressure against the body.

Step back with the right foot and pivot sharply to the right to be next to the opponent. Always pivot in a direction that keeps the weapon away from the opponent.

Straighten the opponent's arm to apply an armbar. The arm should be straight across the torso.

Continue pivoting to the right while pulling back on the opponent's shoulder. This action may break the opponent's arm.

Place the left hand across the opponent's face and apply pressure back and down to take the oppo-

nent to the ground. Pressure applied beneath the nose or on the trachea is equally effective.

Wristlock Technique

Marines use the wristlock technique when an opponent grabs their pistol while it is in the holster with his right hand. To execute the w ristlock technique, Marines—

Grasp the opponent's wrist or hand with the right hand and apply pressure against the body.

Step back and away from the opponent and pivot to the right so the weapon is away from the opponent. Always pivot in a direction that keeps the weapon away from the opponent.

Reach over the opponent's arm with the left hand and grab his hand, applying pressure against his arm with the left forearm. Execute a wristlock.

Incorporate the second hand into the wristlock and, stepping back with the left foot, pivot to the left.

Execute a two-handed wristlock by exerting downward pressure with the thumb and rotating his hand to the left.

Continue to pivot to off-balance the opponent and drive him to the ground.

Note: If the opponent grabs the pistol with his left hand, Marines execute the wristlock with one hand and step in toward the opponent, rather than

away from the opponent.

Softening Techniques

Handgun retention techniques use softening techniques applied to pressure points. Bone pressure

and strikes with the hands (i.e., hammer fist), knees, and feet are also effective softening techniques.

Pressure Points. Marines use pressure point techniques to get the opponent to loosen his grip. Marines use their finger tips to apply pressure to the webbing between the index finger and thumb, the jugular notch, and the brachial plexus tie in. The following figure illustrates pressure applied to the brachial plexus tie in.

Bone Pressure. Bone pressure is the application of pressure on a bone against a hard object to initiate pain compliance. To apply bone pressure, Marines use their hand to trap the opponent's hand on the weapon. Marines apply a slow, steady pressure to the opponent's hand and fingers until his grip is softened or he releases his hold.

Strikes. If it is difficult to apply a retention technique, Marines employ strikes or kicks to force the opponent to loosen his grip. Strikes to the

eyes, the arms (radial nerve), or shoulder (brachial plexus tic in) soften the opponent's grip on the weapon.

Kicks and knee strikes to the peroneal nerve, the femoral nerve, or the groin arc effective because the opponent is typically unprepared to counter the strike.

Stomping on the top of the opponent's foot may distract him or loosen his grip on the weapon.

4. Firearm Disarmament Techniques

Marines use firearm disarmament techniques during a close-range confrontation if they are un-

armed and the opponent has a firearm (pistol). These techniques are equally effective if Marines are armed but do not have time to withdraw and present the weapon. The goal of firearm disarmament techniques is to gain control of the situation

so Marines gain the tactical advantage. The goal is not necessarily to get control of the opponent's weapon.

Pistol to the Front

This technique is used when Marines are unarmed and the opponent has a pistol pointing at their front (e.g., head, chest). The technique is the same if the opponent sticks the pistol under the Marine's chin. To execute the counter when an opponent is pointing a pistol toward the front of a Marine, Marines—

Place the hands close to the weapon, about chest high, palms out.

Use the left hand to grab the opponent's forearm and push the opponent's hand with the pistol away to clear the body from in front of the weapon. At the same time, rotate the right shoulder back to clear the body from the weapon.

Maintain control of the opponent's arm.

Grasp the weapon with the right hand by placing the thumb underneath the pistol and the fingers over top of the pistol.

Keep the right hand wrapped tightly around the muzzle and quickly rotate the pistol in the opponent's hand so the muzzle is facing the opponent.

Grasp and pull the opponent's wrist or forearm away from the body while rotating the weapon.

Rotate the weapon toward the opponent while pulling it up and back and out of the opponent's grasp.

Pistol to the Rear

This technique is performed when Marines are unarmed and the opponent has a pistol pointing to the back of the Marine's head. To execute the counter when the opponent is pointing a pistol to the rear, Marines—

Place the hands close together about chest high, palms out.

Step back with the left foot, pivoting on the right foot so the side is against the opponent's front. This action clears the body from the weapon's line of fire. Keep the left hand up.

Pivot on the left foot to face the opponent and, at the same time, raise the left elbow and reach over the top of the opponent's arm with the left arm.

Wrap the left arm tightly around the opponent's arm above his elbow to control it.

Push on the opponent's shoulder with the right hand while pulling up with the left arm to achieve an armbar. This action releases the opponent's grip on the weapon. If necessary, the opponent can be taken to the ground with a leg sweep.

Note: To execute this technique, the weapon must be close to or touching the rear of the Marine. If the weapon is too far away from the body, this technique would be difficult to execute or it would be ineffective.

CHAPTER 3

HAND-HELD WEAPONS

This chapter describes all techniques for a right-handed person. However, all techniques can be executed from either side.

In drawings, the Marine is depicted in woodland camouflage utilities; the opponent is depicted without camouflage. In photographs, the Marine is depicted in woodland camouflage utilities; the opponent is depicted in desert camouflage utilities.

Marines must know how to defend against attacks when an opponent is either unarmed or armed with a held-held weapon. This chapter addresses the combative use of knives, specific weapons of opportunity, and sticks. However, virtually anything can be used as a hand-held weapon.

1. Fundamentals of Knife Fighting

Marines must be trained in knife fighting techniques. Marines experienced in offensive knife techniques can cause enough damage and massive trauma to stop an opponent. When engaged against each other, experienced knife fighters employ various maneuvers and techniques that are specific to knife fighting. Seldom, if ever, will Marines engage an opponent in a classical knife fight.

Note: When armed with a rifle, Marines are issued a bayonet. When armed with a pistol, Marine are issued a combat knife.

Angles of Attack

There are six angles from which an attack with a knife can be launched:

- Vertical strike coming straight down on an opponent.
- Forward diagonal strike coming in at a 45-degree angle to the opponent.
- Reverse diagonal strike coming in at a 45-degree angle to the opponent.
- Forward horizontal strike coming in parallel to the ground.
- Reverse horizontal strike coming in parallel to the ground.
- Forward thrust coming in a straight line to the opponent.

Target Areas of the Body

During any confrontation, the parts of the opponent's body that are exposed or readily accessible will vary. The goal in a knife fight is to attack the body's soft, vital target areas that are readily accessible (e.g, the face, the sides and front of the neck, the lower abdomen [or groin]).

Neck. Carotid arteries, located on either side of the neck, are good target areas because they are not covered by body armor or natural protection.

Lower Abdomen (or Groin). The lower abdomen (or groin region) is a good target area because it is not covered by body armor.

Heart. The heart, if not covered by body armor, is an excellent target which, if struck, can prove fatal in a matter of seconds or minutes.

Secondary Targets. There are secondary target areas that will cause substantial bleeding if an artery is severed. These target areas are not immediately fatal, but can become fatal if left unattended. Attacks to—

The legs can cause a great deal of trauma and prove fatal. For example, the femoral artery located on the inside of the thigh is a large artery which, if cut, will cause extensive blood loss.

The brachial artery, located between the biceps and triceps on the inside of the arm, can cause extensive bleeding and damage.

The arm's radial and ulnar nerves can cause extensive bleeding and damage.

Movement

Marines can move anywhere within a 360-degree circle around the opponent. This allows accessibility to different target areas of the opponent's body. Marines should avoid being directly in front of an opponent because the opponent can rely on his forward momentum to seize the tactical advantage. If Marines face an opponent, movement is made in a 45-degree angle to either side of the opponent. This angle avoids an opponent's strike and places Marines in the best position to attack an opponent.

Wearing the Combat Knife

Marines must wear the combat knife where it is easily accessible and where it can best be retained. It is recommended the combat knife be worn on the weak side hip, blade down, sharp edge facing forward. Marines can place it behind the magazine pouch where it is easily accessible to them, but not easily grabbed by an opponent.

Grip

The grip on the knife should be natural. Marines grasp the knife's grip with the fingers wrapped around the grip naturally as it is pulled out of its sheath. This is commonly known as a hammer grip. The blade end of the knife is always facing the opponent.

Stance

Marines use the basic warrior stance as the foundation for knife techniques. The left hand is a vertical shield that protects either the ribs or the head and neck. The right elbow is bent with the blade pointing forward toward the opponent' s head. This position serves as an index point, where all techniques are initiated.

Principles of Knife Fighting

The following are key principles of knife fighting:

Execute movements with the knife blade within a box, shoulder-width across from the neck down to the waistline. The opponent has a greater chance of blocking an attack if the blade is brought in a wide, sweeping movement to the opponent.

Close with the opponent, coming straight to the target.

Move with the knife in straight lines.

Point the knife's blade tip forward and toward the opponent.

Apply full body weight and power in each of the knife techniques. Full body weight should be put into the attack in the direction of the blade's movement (slash or thrust).

Apply constant forward pressure with the body and blade to keep the opponent off-balanced.

2. Knife Fighting Techniques

Slashing Techniques

Marines use slashing techniques to close with an enemy. Slashing techniques distract the opponent or damage the opponent so Marines can close in. Typically, Marines target the opponent's limbs, but any portion of the body that is presented can become a target.

Vertical Slash Technique. The vertical slash follows a vertical line straight down through the target. To execute the vertical slash, Marines—

Thrust the right hand out and bring the weapon straight down on the opponent, continuing to drag the knife down through the opponent's body.

Maintain contact on the opponent's body with the blade of the knife.

Forward Slash Technique. The forward slash follows a straight line in a forehand stroke, across the target areas of either the neck (high diagonal slash) or abdominal region (low horizontal slash). To execute the forward slash, Marines-

Extend the right hand while simultaneously rotating the palm up until the knife blade makes contact with the opponent.

Snap or rotate the wrist through the slashing motion to maximize blade's contact with the opponent.

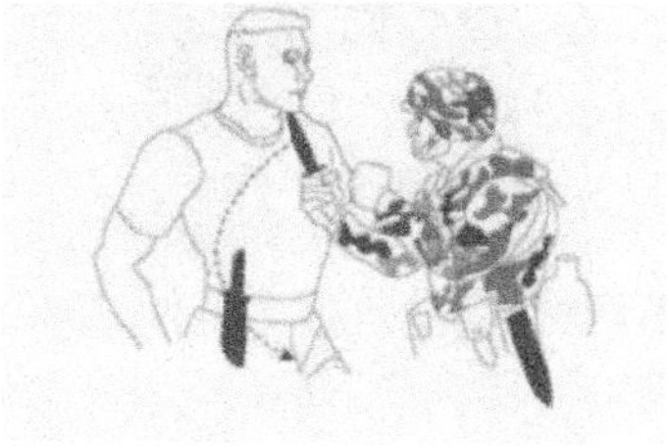

Drag the knife across the opponent's body, from right to left, in a forehand stroke. The movement ends with the forearm against the body and the knife at the left hip with its blade oriented toward the opponent.

Reverse Slash Technique. The reverse slash is a follow-up technique to a forward attack. It allows Marines both a secondary attack and the ability to resume the basic warrior stance. The reverse slash follows a straight line in a backhand stroke, across the target areas of either the neck (high diagonal slash) or abdominal region (low horizontal slash). To execute the reverse slash, Marines—

Extend the right hand while simultaneously rotating the palm down until the knife blade makes contact with the opponent.

Snap or rotate the wrist through the slashing motion to maximize the blade's contact with the opponent.

Drag the knife across the opponent's body, from left to right, in a backhand stroke. Maintain contact on the opponent's body with the blade of the knife.

Thrusting Techniques

The primary objective of knife fighting is to insert the blade into an opponent to cause extensive damage and trauma. This is done with a thrusting technique. Thrusting techniques are more effective than slashing techniques because of the damage they can inflict. However, Marines use slashing techniques to close with the enemy so that they are closer to the opponent, which allows them to use the thrusting technique.

Vertical Thrust. The thrusting motion follows a vertical line straight up through the target (low into the abdomen region or high into the neck). To execute the vertical thrust, Marines—

Thrust the right hand toward the target, inserting the knife blade straight into the opponent.

Pull the knife out of the opponent.

Forward Thrust. The forward thrust follows a straight line straight into the opponent's neck (high thrust) or abdominal region (low thrust). To execute the forward thrust, Marines—

Thrust the right hand, palm down, toward the target, inserting the knife blade straight into the opponent.

Rotate the palm up once the knife is inserted to twist the blade.

Drop the right elbow and bring the knife to the opposite side of the opponent's body from where it was inserted. At the same time, rotate the hips and shoulders downward to bring body weight to bear on the attack.

Reverse Thrust. The reverse thrust is a follow-up technique to a forward attack. It allows Marines both a secondary attack and the ability to resume the basic warrior stance. The reverse thrust follows a horizontal line straight into the opponent's neck (high thrust) or abdominal region (low thrust). To execute the reverse thrust, Marines-

Bend the right arm, crossing the arm to the left side of the body.

Thrust the right hand, palm up, toward the target, inserting the knife blade straight into the opponent.

Rotate the palm down to twist the blade once the knife is inserted.

Bring the knife to the opposite side of the opponent's body from where it was inserted.

3. Weapons of Opportunity

During an unarmed close combat situation, Marines use their bodies as weapons, but they should be ready and able to use anything around them as a weapon. For example, Marines could throw sand or liquid in an opponent's eyes to temporarily impair his vision or smash the opponent's head with a rock or helmet. Marines must use whatever means are available and do whatever it takes to take control of the situation and to win, or they face the possibility of losing their lives. Some weapons of opportunity are discussed in the following subparagraphs.

Entrenching Tool

An entrenching tool (E-tool) is commonly carried by Marines. It can be an excellent weapon, especially when sharpened. Marines can use the E-tool to block, slash, and thrust at an opponent.

Tent Pole and Pins

Marines can use tent poles and pins to block, strike, or thrust at an opponent.

Web Belt

Marines can stretch a web belt between their hands to block attacks by an opponent.

Battlefield Debris

Marines can use debris on the battlefield (e.g., sticks, glass, a sharp piece of metal) to cut, slash, or stab an opponent. They can also use other types of debris such as shovel or ax handles, boards, metal pipes, or broken rifles to strike an opponent or apply a choke.

Helmet

A helmet can be used to strike an opponent on an unprotected area like the head and face. Grasp the rim of the helmet and thrust the arms forward, striking the opponent with the top of the helmet.

4. Fundamentals of Combative Stick

On the battlefield, Marines must be ready and able to use anything as a weapon. They must learn and be able to use techniques that can be employed with most weapons of opportunity. Among these techniques are combative stick techniques. Combative stick techniques can be used with a stick, a club, a broken rifle, an E-tool, or even a web belt.

Angles of Attack

There are six angles from which an attack with a hand-held weapon can be launched:

Vertical strike coming straight down on an opponent.
Forward diagonal strike coming in at a 45-degree angle to the opponent.
Reverse diagonal strike coming in at a 45-degree angle to the opponent.
Forward horizontal strike coming in parallel to the ground.
Reverse horizontal strike coming in parallel to the ground.
Forward thrust coming in a straight line to the opponent.

Grip

Grasp the stick about 2 inches from its base.

Stance

The basic warrior stance serves as the foundation for combative stick techniques. The left hand becomes a vertical shield that protects the ribs or the head and neck. Depending on how heavy the weapon is, it should be held at a level approximately shoulder height.

Movement

Movement during combative stick techniques is the same as it is for other close combat techniques. Marines can move anywhere within a 360-dcgrcc circle around the opponent. This allows accessibility to different target areas of the opponent's body and gains a tactical advantage. Marines should avoid being directly in front of an opponent because the opponent can rely on his forward momentum to seize the tactical advantage. If Marines face an opponent, movement is made in a 45-dcgrcc angle to cither side of the opponent. This angle avoids an opponent's strike and places Marines in the best position to attack an opponent.

5. Combative Stick Techniques

Strikes

Strikes are intended to inflict as much damage on an opponent as possible. Striking techniques apply to a weapon of opportunity such as a stick, a tent pole, a club, a broken rifle, an E-tool, or a Pipe-

Vertical Strike. To execute the vertical strike, Marines—

Bend the right arm, extending the weapon over the back of the right shoulder.

Rotate the forearm straight down off the elbow to bring the weapon down on the opponent.

Rotate the hips and shoulders forcefully toward the opponent.

Follow through with the strike by allowing the weight of the weapon to go through the target area of the body.

Forward Strike. To execute a forward strike, Marines—

Step forward with the left foot in the direction of the strike.

Bend the right arm with the elbow extending out to the right and the weapon extended over the right shoulder.

Rotate the forearm to the right of the elbow to bring the weapon down onto the opponent. At the same time, forcefully rotate the hips and shoulders toward the opponent.

Follow through with the strike by allowing the weight of the weapon to go through the target area of the body.

Reverse Strike. The reverse strike is a follow-up technique to a forward strike. It allows Marines both a secondary attack and the ability to resume the basic warrior stance.To execute a reverse strike, Marines—

Step forward with the right foot in the direction of the strike.

Bend the right arm with the hand near the left shoulder. The weapon is extended over the left shoulder.

Rotate the forearm to the right of the elbow to bring the weapon down onto the opponent. At the same time, forcefully rotate the hips and shoulders toward the opponent.

Follow through with the strike by allowing the weight of the weapon to go through the target area of the body.

Forward Thrust. To execute the forward thrust, Marines—

Grasp the stick with the left hand, palm up, in a position where the stick can be controlled with two hands.

Lift the left leg and lunge forward off the ball of the right foot. At the same time, thrust the end of the weapon directly toward the opponent by thrusting both hands forward in a straight line.

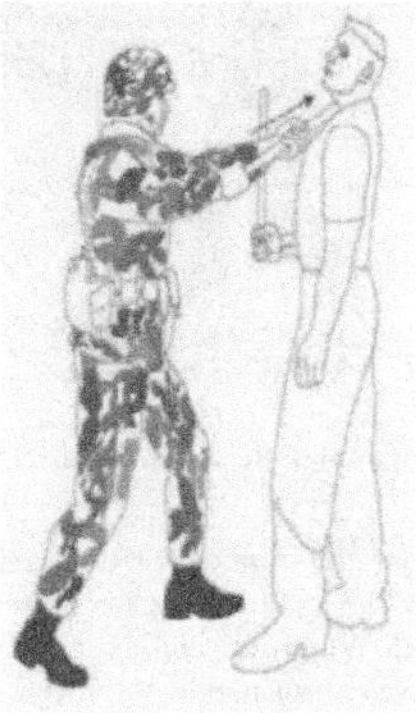

6. Blocking Techniques

A block is meant to deter or deflect an attack by an opponent. A block sets up Marines for a follow-on attack against the opponent. Blocks are executed by deflecting, rather than hitting or following through like a strike.

Blocks Against Unarmed Attacks

To block against an unarmed attack, Marines—

Step forward at a 45-degree angle with either the right or left foot. This moves the body out of the line of attack.

Raise the left arm and block the strike with the meaty portion of the forearm.

Employ the stick by one of two techniques:

Strike the opponent with the stick.
Use the stick to block on two points of contact. When the stick is used to block, it serves as an incidental strike. This technique is only used if Marines have closed with the opponent and are inside his strike.

Blocks Against Armed Attacks

Block for a Vertical Strike. To execute the block for a vertical strike, Marines—

Step forward with the left foot at a 45-degree angle to the left. This moves the body out of the line of attack.

Block on two points of contact to disperse the impact of the attack:

Block the opponent's stick by positioning the stick so it is perpendicular to the opponent' s stick. If the stick is not perpendicular to the opponent's stick, the stick can slide through and make contact on the Marine.
Block the opponent's wrist or forearm with the back of the left forearm.

Note: Use the stick to block the opponent's arm if closer to the opponent. It is the same movement as blocking with the arm, except the opponent's arm is blocked with both the stick and the arm.

Block for a Forward Strike. To execute the block for a forward strike, Marines-
Step forward with the left foot at a 45-degree angle to the left. This moves the body out of the line of attack and inside the opponent's strike.

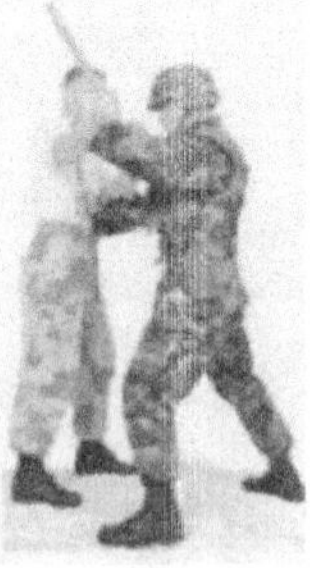

Block on two points of contact to disperse the impact of the attack:

- Block the opponent's wrist or forearm with the meaty portion of the back of the left forearm.
- Strike the opponent's attacking biceps with the stick.

Block for a Reverse Strike. To execute the block for a reverse strike, Marines—

Step forward at a 45-degree angle with the right foot to the right. This moves the body out of the line of attack and inside the opponent's strike.

Block on two points of contact to disperse the impact of the attack:

Block the opponent's stick by positioning the stick so it is perpendicular to the opponent's stick.

Block the opponent's forearm with the meaty portion of the left forearm.

Note: If closer to the opponent, block his triceps with the back of the left forearm and strike his forearm with the stick.

7. Unarmed Against Hand-Held Weapons

If Marines are engaged against an opponent with a knife, a stick, or some other weapon of opportunity, they must establish and maintain an offensive mindset, not a defensive mindset. Their survival depends on it. They cannot afford to think about getting cut or hurt.

Angles of Attack

Before Marines can learn to block or counter an attack with a hand-held weapon (e.g., knife, stick), they must know from what angle the opponent is attacking. There arc six angles from which an opponent will typically attack with a hand-held weapon:

Vertical strike coming in straight down on the Marine.
Forward diagonal strike coming in at a 45-degree angle to the Marine.
Reverse diagonal strike coming in at a 45-degree angle to the Marine.
Forward horizontal strike coming in parallel to the ground.
Reverse horizontal strike coming in parallel to the ground.
Forward thrust coming in a straight line to the Marine.

Blocks

If the opponent has a hand-held weapon (e.g., knife or stick), Marines parry the opponent's hand or arm to block the attack.

Basic Block. To execute the basic block technique, Marines-

Step forward at a 45-degree angle to move out of the line of the attack. Always step in the direction of the strike.

Thrust the forearms forward, hands up, against the opponent's attacking arm. Contact is made on the opponent's arm with the backs of the fore-arms.

Block for a Vertical Strike. To execute the block against a vertical strike, Marines—

Step forward with the left foot at a 45-degree angle to the left to move out of the line of attack.

Thrust the forearms forward, hands up, against the outside of the opponent's attacking arm.

Block for a Forward Strike. To execute the block against a forward diagonal or forward horizontal strike, Marines-

Step forward with the left fool inside the opponent's attacking arm.

Block the attack with both arms bent so the fore-arms make contact with the opponent's biceps and forearm.

Block for a Reverse Strike. To execute the block against a reverse diagonal or reverse horizontal strike, Marines—

Step forward with the right foot to the outside of the opponent's attacking arm.

Block the attack with both arms bent so the fore-arms make contact with the opponent's triceps and forearm.

Block for a Forward Thrust. To execute the block against a forward thrust, Marines—

Bend at the waist, move the hips backward, and jump backward with both feet to move away from the attack. This action is known as "hollowing out."

Hollow out and block the attack with the arms bent and hands together on top of the attacking arm.

Overlap the hands slightly so one thumb is on top of the other hand's index finger. The other thumb should be under the other hand's index finger.

8. Counters to Hand-Held Weapon Attacks

Principles of Counters

A counter is used to control the situation to regain the tactical advantage and end the fight. Regardless of the type of weapon or angle of attack, the following principles apply to countering the attack with a hand-held weapon:

Move out of the line of attack. Movement is executed in a 45-degree angle forward to the left or right.

i Block the attack.

Note: The first two actions are taken simultaneously.

i Take control of the weapon by controlling the hand or arm that is holding the weapon. Never attempt to grab the opponent's weapon.

i Execute the appropriate follow-up to end the fight; e.g., strikes, joint manipulations, throws, or takedowns (see chap. 8). Marines should continue to attack the opponent until the fight ends.

Counter Techniques

There are two techniques that can be used to counter any armed attack: forward armbar counter and reverse armbar counter. These techniques can be used to counter a vertical attack, a forward diagonal strike, or a forward horizontal strike. With minor variations, the same techniques arc used to counter reverse strikes. A third technique, the bent armbar counter, is used to counter a vertical attack.

Forward Armbar Counter. To execute the forward armbar counter to an attack coming from a forward strike, Marines-

Step forward with the left foot inside the opponent' s attacking arm.

Block the attack with both arms bent so the forearms make contact with the opponent's biceps and forearm.

Slide the left arm over the opponent's forearm and wrap the arm tightly around his arm, trapping the opponent's attacking arm between the biceps and torso.

Place the right hand on the opponent's shoulder or upper arm to further control his arm and to effect an armbar.

Execute an armbar and continue to exert steady pressure against the arm.

Reverse Armbar Counter. To execute the reverse armbar counter to an attack coming from a forward strike, Marines-

Step forward with the left foot inside the opponent' s attacking arm.

Block the attack with both arms bent so the forearms make contact with the opponent's biceps and forearm.

Control the opponent's arm with the left hand and pivot to the right so the back is against the opponent's side. Immediately slide the right arm over the opponent's biceps and wrap the arm tightly around his arm, trapping the opponent's attacking arm between the biceps and torso.

Grasp the opponent's wrist with the left hand and twist his thumb away from the body.

Take the opponent to the ground with an armbar takedown if he does not drop the weapon from the wristlock.

Bent Armbar Counter. This counter is particularly effective against a vertical attack. To execute the bent armbar counter, Marines—

Step forward with the left foot inside the opponent's attacking arm.

Block the attack with both arms bent so the forearms make contact with the opponent's biceps and forearm.

Grasp the opponent's forearm with the left hand. At the same time, slide the right arm underneath the opponent's triceps and grasp the opponent's forearm or wrist with the right hand.

Apply downward pressure with the hands against the opponent's forearm to off-balance the opponent.

(reverse blank)

CHAPTER 4

STRIKES

This chapter describes all techniquesfor a right-lmndedperson. However, all techniques can be executed from either side.

In drawings, the Marine is depicted in woodland camouflage utilities; the opponent is depicted without camouflage. In photographs, the Marine is depicted in woodland camouflage utilities; the opponent is depicted in desert camouflage utilities.

Strikes are unarmed individual hitting techniques. Strikes use the hands, elbows, knees, feet, and, in some instances, other parts of the body as personal weapons. Marines must know how to execute strikes effectively. They must also know how to counter strikes from an opponent.

1. Principles of Punches

Muscle Relaxation

Muscle relaxation is crucial when executing punches. The natural tendency in a fight is to tense up, which results in rapid fatigue and decreased power generation. Marines who remain relaxed during a close combat situation generate greater speed, which results in greater generation of power. Relaxing the forearms generates speed and improves reaction time. At the point of impact, Marines clench the fist to cause damage to the opponent and avoid injury to the wrist and hand.

Weight Transfer

Weight transfer is necessary to generate power in a punch. Marines accomplish this by-

Rotating their hips and shoulders into the attack.

Moving their body mass straight forward or backward in a straight line.

Dropping their body weight into an opponent. Body mass can be transferred into an attack from high to low or from low to high.

Rapid Retraction

When Marines deliver a punch, rapid retraction of the fist is important. Once the hand has made contact with the target, Marines quickly return to the basic warrior stance. Rapid retraction-

Returns the hand and arm to the protection afforded by the basic warrior stance.

Prevents the opponent from being able to grab the hand or arm.

Permits the hand and arm to be "chambered" or "re-cocked" in preparation for delivering a subsequent punch.

Telegraphing

Telegraphing a strike occurs when body movements inform the opponent of the intention to launch an attack. Staying relaxed helps to reduce telegraphing.

Often, an untrained fighter telegraphs his intention to attack by drawing his hand back in view of his opponent, changing facial expression, tensing neck muscles, or twitching. These movements, however small, immediately indicate an attack is about to be delivered. If the opponent is a trained fighter, he may be able to evade or counter the attack. If the opponent is an untrained fighter, he may still be able to minimize the effect of an attack.

2. Punches

Punches may be thrown during any hand-to-hand confrontation. Most people resort to punching because it is a natural reaction to a threat. The purpose of a punch is to stun the opponent or to set him up for a follow-up finishing technique. However, punches should only be executed to the soft tissue areas of an opponent. A correctly delivered punch maximizes the damage to an opponent while minimizing the risk of injury to Marines.

Basic Fist

Punches are executed using the basic fist. To make the basic fist, the fingers are curled naturally into the palm of the hand and the thumb is placed across the index and middle fingers. Do not clench the fist until movement has begun. This reduces muscular tension in the forearm and increases speed and reaction time. Just before impact, Marines exert muscular tension on the hand and forearm to maximize damage to the opponent and reduce their chances of injury. Contact should be made with the knuckles of the index and middle fingers.

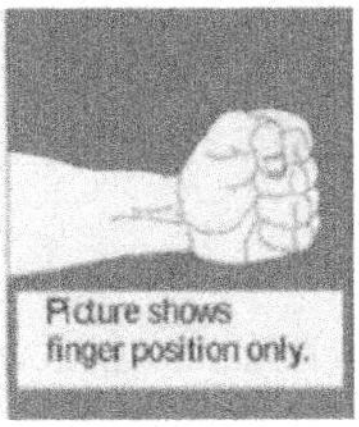

When striking with the basic fist, Marines must keep the hand straight, or in line, with the wrist to avoid injury to the wrist.

Lead Hand Punch

The lead hand punch is a snapping straight punch executed by the forward or lead hand. It is a fast punch designed to keep the opponent away and to set up an attack. A lead hand punch conceals movement and allows Marines to get close to the opponent. Lead hand punches should strike soft tissue areas, if possible. To execute the lead hand punch, Marines-

Snap the lead hand out to nearly full extension, while rotating the palm to the ground.

Contact the opponent with the first two knuckles of the fist.

Retract the hand immediately, resuming the basic warrior stance.

Rear Hand Punch

The rear hand punch is a snapping punch executed by the rear (right) hand. It is a power punch designed to inflict maximum damage on the opponent. Its power comes from pushing off the rear leg and rotating the hips and shoulders. To execute the rear hand punch, Marines—

Rotate the hips and shoulders forcefully toward the opponent and thrust the rear hand straight out, rotating the palm down, to nearly full extension.

Shift body weight to the lead foot while pushing off on the ball of the rear foot.

Contact the opponent with the first two knuckles of the fist.

Retract the hand immediately.

Uppercut

The uppercut is a powerful punch originating below the opponent's line of vision. It is executed in an upward motion traveling up the centerline of the opponent's body. It is delivered in close and usually follows a preparatory strike that leaves the target area unprotected. When delivered to the chin or jaw, the uppercut can render an opponent unconscious, cause extensive damage to the neck, or sever the tongue. To execute the uppercut, Marines—

Bend the arms, rotating the palm inboard. The distance the arms bend depends on how close the opponent is.

Rotate the hips and shoulders forcefully toward the opponent, thrusting the fist straight up toward the opponent's chin or jaw.

Contact the opponent with the first two knuckles of the fist.

Retract the hand immediately.

Hook

The hook is a powerful punch that is executed close in and is usually preceded by a preparatory strike. To execute the hook, Marines—

Thrust the right arm in a hooking motion toward the opponent, keeping the elbow bent while forcefully rotating the right shoulder and hip toward the opponent.

Contact the opponent with the first two knuckles of the fist. Continue rotating the shoulder and hip, following through with the fist to the target.

Retract the hand immediately.

3. Strikes with the Upper Body

Strikes stun the opponent or set him up for a follow-up finishing technique. The hands, forearms, and elbows are individual weapons of the arms that can be used to execute strikes including the hammer fist, chin jab, knife hand, eye gouge, and elbow strikes. These strikes provide a variety of techniques that can be used in any type of close combat situation.

Principles of Execution

There are several principles of execution that ensure a strike's effectiveness.

Generating Power. Weight transfer is necessary to generate maximum power in a punch. Marines accomplish this by-

Rotating the hips and shoulders into the attack.
Moving the body mass straight forward or backward in a straight line.
Dropping body weight into an opponent.
Body mass can be transferred into an attack from high to low or from low to high.

Muscular Tension. The arms are relaxed until the moment of impact. At the point of impact, Marines apply muscular tension in the hand and forearm to maximize damage to the opponent and to avoid injury to the hand. The arms are relaxed until the moment of impact.

Hit and Stick and Follow-Through. A strike should be delivered so that the weapon (e.g., hand, elbow) hits and remains on the impact site (target) and follows through the target. This technique inflicts maximum damage on the opponent.

Strikes with the arms are executed with "heavy hands"; i.e., the strike is executed by driving through with the strike to allow the weight of the hand to go through the target area of the body. Contact with the opponent should be made with the arm slightly bent, and the arm extends as it moves through the target. This technique allows Marines to deliver strikes effectively without executing full force.

Movement. Movement puts Marines in the proper position for launching an attack against an opponent as well as providing protection.

Movement is initiated from the basic warrior stance and ends by resuming the basic warrior stance. Strikes can be performed with either the left or right arm depending upon—

The angle of attack.
The position of the opponent.
The opponent's available, vulnerable target areas.

Target Areas of the Body

For each strike, there are target areas of the body that, when struck, maximize damage to an opponent. Strikes use gross motor skills as opposed to fine motor skills. The target areas of the body are just that—areas. Pinpoint accuracy on a specific nerve is not needed for the strike to be effective.

Hammer Fist

Striking with the hammer fist concentrates power in a small part of the hand, which, when transferred to the target, can have a devastating effect. The striking surface of the hammer fist is the meaty portion of the hand below the little finger. To execute the hammer fist strike, Marines-

Make a fist and bend the arm at approximately a 45- to 90-degree angle. At the same time, rotate the right hip and right shoulder backward.

Thrust the fist forward onto the opponent while rotating the right hip and shoulder forward.

Rotate the wrist so the hammer fist makes contact on the opponent.

Chin Jab

The chin jab can immediately render an opponent unconscious and cause extensive damage to the neck and spine. The striking surface is the heel of the palm of the hand. To execute the chin jab, Marines—

Bend the right wrist back at a 90-degree angle with the palm facing the opponent and the fingers pointing up.

Keep the right arm bent and close to the body. Extend the hand into a concave position with the fingers slightly spread apart.

Step forward with the left foot toward the opponent, keeping the feet approximately shoulder-width apart and the knees bent. This is done to close with the opponent.

Keep the right arm bent and close to the side. Thrust the palm of the hand directly up under the opponent's chin. At the same time, rotate the right hip forward to drive the body weight into the attack to increase the power of the strike. The attack should travel up the centerline of the opponent's chest to his chin.

Knife Hand

The knife hand is one of the most versatile and devastating strikes. The striking surface is the cutting edge of the hand, which is the meaty portion of the hand below the little finger extending to the top of the wrist. The striking surface is narrow, allowing strikes on the neck between the opponent's body armor and helmet. The knife hand strike is executed from one of three angles: outside, inside, and vertical.

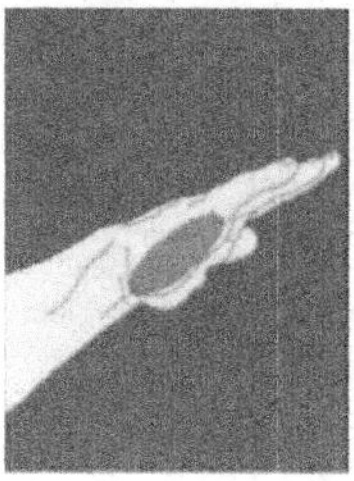

Outside Knife Hand. To execute the outside knife hand strike, Marines—

Execute a knife hand by extending and joining the fingers of the right hand and placing the thumb next to the forefinger (like saluting).

Retract the right hand. At the same time, rotate the right hip and right shoulder backward.

Thrust the knife hand forward (horizontally) onto the opponent while rotating the right hip and shoulder forward.

Inside Knife Hand. To execute the inside knife hand strike, Marines—

Execute a knife hand.

Bring the right hand over the left shoulder. At the same time, rotate the right shoulder forward and the left hip forward.

Thrust the knife hand forward (horizontally) onto the opponent while rotating the right hip and shoulder forward and the left shoulder backward.

Vertical Knife Hand. When thrown vertically, the knife hand strike comes straight down in a straight line.

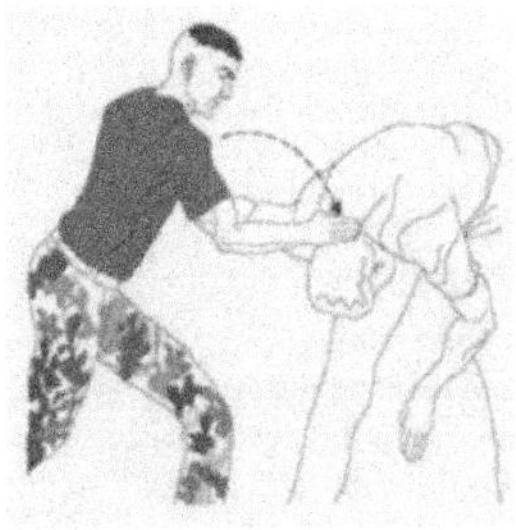

Eye Gouge

The eye gouge is used to attack an opponent's eyes, blinding him so follow-up strikes can be executed. The striking surface is the tips of the fingers or thumb. To execute the eye gouge, Marines—

Extend the right hand with the fingers slightly spread apart to allow entry into the eye sockets.

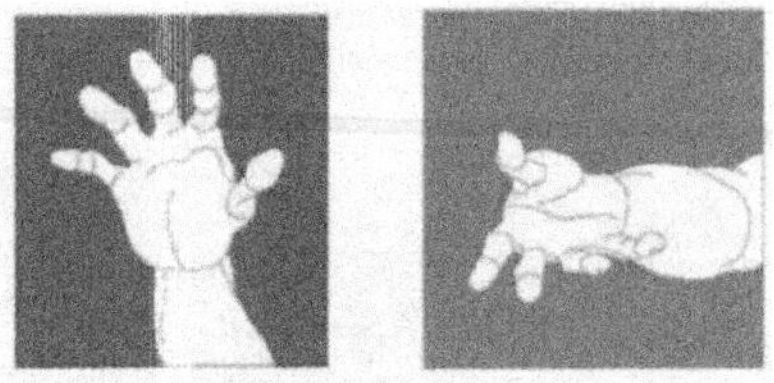

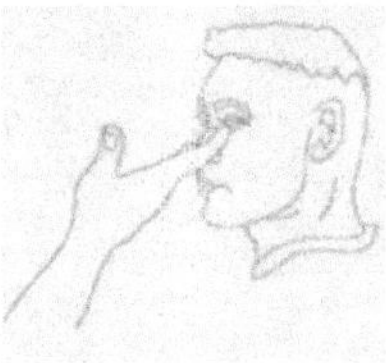

Place the palm of the hand either toward the ground or toward the sky and thrust the right hand forward into the opponent's eyes.

Thrust the hand forward at the opponent's nose level so the fingers or thumb slide naturally into the grooves of the opponent's eye sockets.

Elbow Strikes

The elbow is a powerful weapon that can be used in several different ways to attack virtually any part of an opponent's body. Elbow strikes can be performed either vertically (upward or downward) or horizontally (forward or reverse). The striking surface is 2 inches above or below the point of the elbow, depending upon the angle of attack, the opponent's attack angle, and the position of the opponent.

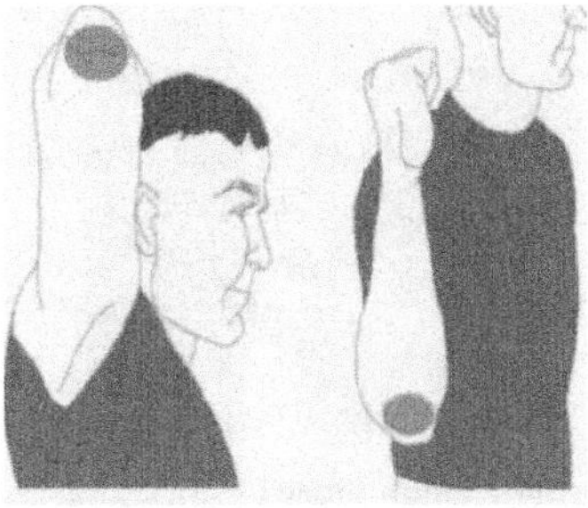

Above Elbow Below Elbow

Vertical Elbow Strike (Up). To execute an upward vertical elbow strike, Marines-

Bend the right elbow, keeping the fist close to the body. The fist is at shoulder level and the elbow is next to the torso.

Thrust the elbow vertically upward toward the opponent while rotating the right shoulder and hip forward to generate additional power.

Contact the opponent with the right forearm 2 inches above the point of the elbow.

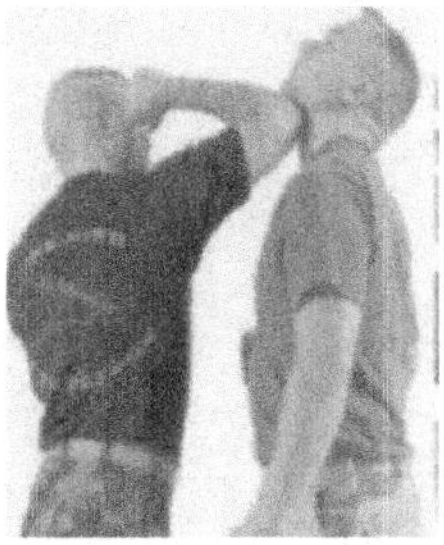

Vertical Elbow Strike (Down). To execute a downward vertical elbow strike, Marines—

Bend the right elbow, keeping the fist close to the body. The fist is on the shoulder and the elbow is raised well above the shoulder.

Thrust the elbow vertically downward toward the opponent while dropping body weight into the attack to generate additional power.

Contact the opponent with the right triceps 2 inches above the point of the elbow.

Horizontal Elbow Strike (Forward). To execute the forward horizontal elbow strike, Marines-

Tuck the right fist near the chest with the palm heel facing the ground.

Thrust the right elbow horizontally forward toward the opponent. The forearm is parallel to the ground.

Rotate the right shoulder and hip forward to generate additional power.

Contact the opponent with the right forearm 2 inches below the point of the elbow.

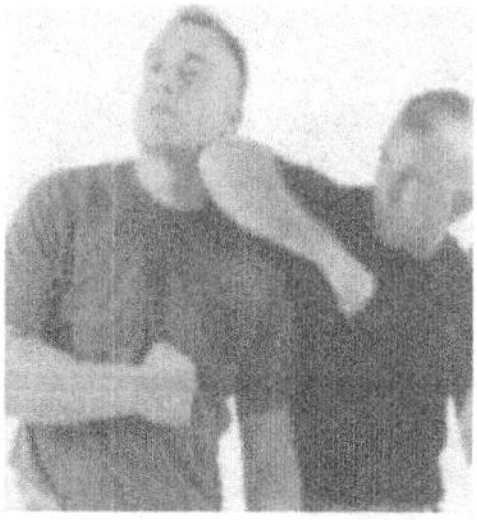

Horizontal Elbow Strike (Rear). To execute the rear horizontal elbow strike, Marines-

Tuck the right fist near the left shoulder with the palm heel facing the ground. At the same time, rotate the right shoulder forward and the left hip forward.

Thrust the right elbow horizontally rearward toward the opponent. The forearm is parallel to the ground and the hand moves toward the direction of the attack.

Rotate the right hip back and the right shoulder backward to generate additional power.

Contact the opponent with the right triceps 2 inches above the point of the elbow.

4. Strikes with the Lower Body

The legs are the body's most powerful weapons because they use the largest muscle groups to generate a strike. Legs are also less prone to injury. The feet are the preferred choice for striking because they are protected by boots. Marines use their feet, heels, and knees to execute kicks, knee strikes, and stomps.

Kicks

The purpose of a kick is to stop an opponent's attack or to create an opening in his defense in order to launch an attack. Kicks can be performed with the left (lead) leg or the right (rear) leg. Kicks with the rear leg have greater power because the hips are rotated into the attack. However, the rear leg is further away from the opponent so a strike with the rear leg will not contact the opponent as quickly as a strike with the lead leg.

Front Kick. The front kick is executed when the opponent is in front of the Marine. The front kick, delivered with either the rear or lead leg, is effective for striking below the waist. Attempting to kick higher results in diminished balance and provides the enemy with a greater opportunity to grab the leg or foot. The striking surfaces arc the toe of the boot or the bootlaces. To execute the front kick, Marines-

Raise the left knee waist high, pivot the hips into the attack, and thrust the left foot forward toward the opponent.

Contact the opponent with the toe of the left boot or bootlaces.

Return to the basic warrior stance.

Side Kick. The side kick, delivered with the lead leg, is effective for striking the knees. The side kick is executed when the opponent is to the side of the Marine. The striking surface is the outside cutting edge of the boot near the heel. To execute the side kick, Marines—

Raise the right knee waist high and rotate the right hip forward.

Thrust the right foot to the right side toward the opponent, turning the foot at a 90-degree angle to maximize the striking surface on the opponent.

Contact the opponent with the cutting edge of the right boot.

Return to the basic warrior stance.

Knee Strikes

Knee strikes are excellent weapons during the close range of close combat fighting. The knee strike is generally delivered in close.

Vertical Knee Strike. The striking surface is the thigh, slightly above the knee. To execute the vertical knee strike, Marines-

Raise the right knee and drive it up forcefully into the opponent. Power is generated by thrusting the leg upward.

Contact the opponent 2 inches above the right knee.

Horizontal Knee Strike. The horizontal knee strike is executed with the leg generally parallel with the ground while rotating the hips to generate power. It is often delivered to the peroneal nerve. The striking surface is the front of the leg, slightly above or below the knee. To execute the horizontal knee strike, Marines-

Raise the right knee, rotate the right hip forward while pivoting on the left foot, and drive the knee horizontally into the opponent.

Contact the opponent 2 inches above the right knee.

Stomps

Stomps are delivered with the feet, usually when the opponent is down. Remember, when the opponent is down, Marines take whatever target is available.

Vertical Stomp. The vertical stomp allows Marines to remain upright and balanced, to rapidly deliver multiple blows with either foot, and to quickly and accurately attack the target. It is the preferred stomp. The striking surface is the flat bottom of the boot or the cutting edge of the heel. To execute the vertical stomp, Marines—

Raise the right knee above the waist with the right leg bent at approximately a 90-degree angle.

Drive the flat bottom of the right boot or the cutting edge of the right heel down onto the opponent forcefully. At the same time, bend the left knee slightly to drop the body weight into the strike.

Ax Stomp. The striking surface of the ax stomp is the cutting edge of the heel. To execute the ax stomp, Marines-

Raise the right heel above the waist, keeping the right leg straight.

Drive the cutting edge of the right heel down onto the opponent forcefully. At the same time, bend the left knee slightly to drop the body weight into the strike.

5. Counters to Strikes

In a close combat situation, an opponent will attempt to strike Marines with punches and kicks. When an opponent uses a strike, Marines must first avoid the strike. This is accomplished by moving quickly and blocking. Next, Marines must get into an offensive position. This allows Marines to use offensive strikes to attack the opponent. Regardless of the strike, the counter to a strike requires Marines to move, block, and strike.

Move

The first step in countering a strike is to move out of the way of the strike's impact. Movement removes Marines from the opponent's intended strike point and positions Marines to attack. Movement is executed at approximately a 45-degree angle to the front or rear. Movement is always initiated from the basic warrior stance. Return to the basic warrior stance with the toe of the lead foot pointing toward the opponent once the movement is complete.

Block

Different blocks are executed based on the strike. These will be covered by individual counters.

Strike

Any of the upper body or lower body strikes can be executed as a follow-on attack or part of the counter to an opponent's strike. The follow-on strike is determined by the angle to the opponent, the position of the opponent, and the opponent's available vulnerable target areas.

Counters to Punches

Counter to a Lead Hand Punch. This counter is used when the opponent throws a lead hand punch. To execute the counter to the lead hand punch, Marines-

Step forward and to the left at approximately a 45-degree angle, moving in to the outside of the opponent's attacking arm.

Raise the left arm and block or deflect the opponent's lead hand with the palm of the hand or the meaty portion of the forearm.

Hit and stick by leaving the left arm against the opponent's right arm while stepping forward and to the left at approximately a 45-degree angle to close with the opponent.

Execute a finishing technique, such as a strike or a kick, to the opponent's exposed target areas.

Counter to a Rear Hand Punch. This counter is used when the opponent throws a rear hand punch. To execute the counter to the rear hand punch, Marines—

Step forward and to the left at approximately a 45-degree angle, moving in to the outside of the opponent's attacking arm.

Raise the left arm and block or deflect the opponent's rear hand with the palm of the hand or the meaty portion of the forearm.

Hit and stick by leaving the left arm against the opponent's right arm while stepping forward and to the right at approximately a 45-degree angle to close with the opponent.

Execute a finishing technique, such as a strike or a kick, to the opponent's exposed target areas.

Counters to Kicks

Counter to a Front Kick (Left or Lead Leg). This counter is used when the opponent executes a front kick with his left leg. To execute the counter to a front kick, Marines-

Step forward and to the right at approximately a 45-degree angle, moving in to the outside of the opponent's striking leg.

Lower the left arm and block or deflect the opponent's leg with the palm of the hand or the meaty portion of the forearm.

Hit and stick by leaving the left arm against the opponent's leg while stepping forward and to the left at approximately a 45-degree angle to close with the opponent.

Execute a finishing technique, such as a strike or a kick, to the opponent's exposed target areas.

Counter to a Front Kick (Right or Rear Leg). This counter is used when the opponent executes a front kick with his right leg. To execute the counter to a front kick, Marines—

Step forward and to the left at approximately a 45-degree angle, moving in to the outside of the opponent's striking leg.

Lower the left arm and block or deflect the opponent's leg with the palm of the hand or the meaty portion of the forearm.

Hit and stick by leaving the left arm against the opponent's leg while stepping forward and to the right at approximately a 45-degree angle to close with the opponent.

CHAPTER 5

THROWS

This chapter describes all techniques for a right-handed person. However, all techniques can be executed from either side.

The Marine is depicted in camouflage utilities. The opponent is depicted without camouflage.

Marines use throwing techniques to maintain the tactical advantage and to throw the opponent to the ground during close combat. Throws apply the principles of balance, leverage, timing, and body position to upset an opponent's balance and to gain control by forcing the opponent to the ground. Throwing techniques are effective because they are size- and gender-neutral, and they rely on the momentum and power generated by the opponent rather than the strength or size of the Marine. Marines also execute a throw as a devastating attack against an opponent, possibly causing unconsciousness or broken limbs. When Marines execute throws, they must maintain balance and, simultaneously, prevent the opponent from countering a throw or escaping after being forced to the ground.

1. Turning Throw

Marines use a turning throw to take the opponent to the ground while they remain standing. A turning throw can also be executed from a stationary position. It is particularly effective if Marines and the opponent are wearing gear. To execute the turning throw, Marines-

Grasp the opponent's right wrist with the left hand.

Step forward with the right foot, place it against the outside of the opponent's right foot, and pivot so the back of the heel is next to the middle of the opponent's foot.

Hook the opponent's right arm with the right arm and pinch his arm between the biceps and forearm, touching the opponent with the body.

Pull the opponent's wrist downward, keeping it close to the body.

Pivot to the left on the ball of the foot and continue pulling downward on the opponent's wrist while rotating the wrist outward to off-balance him.

2. Hip Throw

Marines use a hip throw to take the opponent to the ground while they remain standing. A hip throw is particularly effective if the opponent is moving forward or pushing. Marines use the opponent' s forward momentum to execute the hip throw. To execute the hip throw, Marines-

Grasp the opponent's right wrist with the left hand.

Step forward with the right foot and place it against the outside of the opponent's right foot.

The back of the heel should be next to the opponent' s foot.

Step back with the left foot and rotate on the ball of the foot. The back of the heel is next to the opponent' s toe. The knees are bent.

Rotate at the waist and hook the right hand around the back of the opponent's body (anywhere from his waist to his head). The side and hip should be against the opponent.

Rotate the hip against the opponent. The hips must be lower than the opponent's.

Use the right hand to pull the opponent up on the hip to maximize contact.

Pull the opponent's arm across the body and, at the same time, lift the opponent off the ground slightly by bending at the waist, straightening the legs, and rotating the body to the left.

3. Leg Sweep

Marines use a leg sweep to take the opponent to the ground while they remain standing. A leg sweep is particularly effective if the opponent is already off-balanced and moving backward or pulling on the Marine. To execute the leg sweep, Marines—

Grab the opponent's right wrist with the left hand and grab the opponent' s left shoulder with the right hand.

Note: Grab the opponent's clothing or gear if his wrist and shoulder cannot be grabbed.

Step forward with the left foot and place it on the outside of the opponent's right foot.

Pull the opponent's wrist downward, close to the body, and push his shoulder backward to off-balance him.

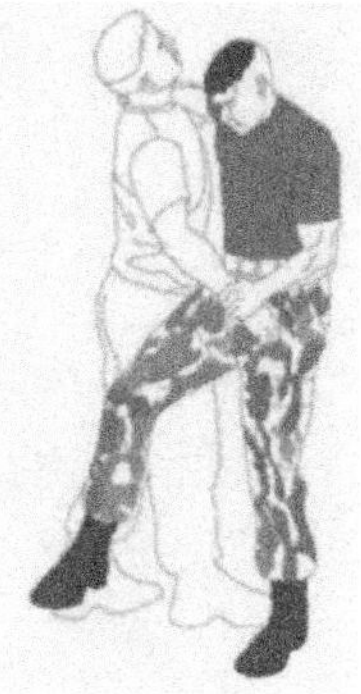

Raise the right knee no higher than the waist.

Kick the foot past the opponent's right leg.

Use the heel of the boot to make contact with the opponent's calf (anywhere from the top of the calf down to the Achilles tendon or on the inside of the calf).

Sweep the opponent to the ground.

CHAPTER 6

CHOKES AND HOLDS

This chapter describes all techniques for a right-handed person. However, all techniques can be executed from either side.

The Marine is depicted in camouflage utilities. The opponent is depicted without camouflage.

WARNING

During training, never execute a choke at full force or full speed and never hold a choke for more than 5 seconds.

When Marines correctly perform a choke, they render an opponent unconscious in as little as 8 to 13 seconds. Chokes are easily performed regardless of size or gender. Marines must know how to apply chokes and how to counter a choke or a hold executed by an opponent.

1. Types of Chokes

There are two types of chokes: an air choke and a blood choke. An air choke closes off the airway to the lungs, thereby preventing oxygen from reaching the heart. A blood choke cuts off the blood flow to the brain. Both types can result in unconsciousness and eventual death for an opponent.

Air Choke

An air choke is performed on the opponent's windpipe (or trachea), cutting off the air to the lungs and heart. If Marines execute the air choke properly, the opponent loses consciousness within 2 to 3 minutes. Due to the length of time it takes to immobilize the opponent, air chokes are not recommended.

Blood Choke

A blood choke is performed on the opponent's carotid artery, which carries oxygen-enriched blood from the heart to the brain. The carotid artery is located on both sides of the neck. If Marines execute a blood choke properly, the opponent will lose consciousness within 8 to 13 seconds. The blood choke is the preferred choke because its intended effect (i.e., the opponent losing consciousness) can be executed quickly, ending the fight.

2. Chokes

Front Choke

Marines execute a front choke when they are facing the opponent. Marines use the opponent's lapels or collar to execute a front choke. To execute the front choke, Marines-

Grab the opponent's right lapel with the right hand, making certain that the knuckles or the back of the hand are against the opponent's carotid artery on the right side of his neck.

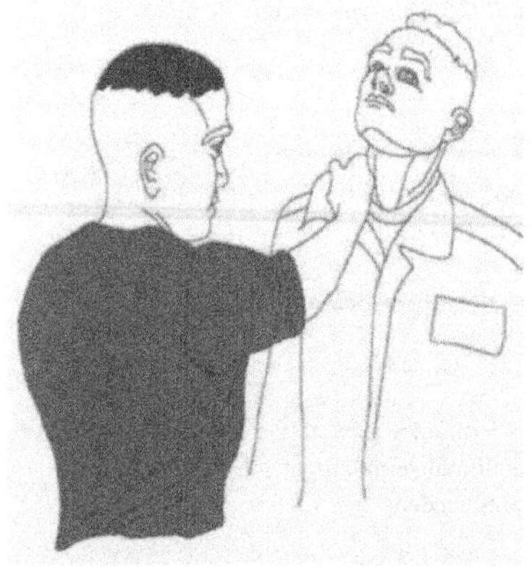

Keep the right hand against the opponent's neck, reach under the opponent's right arm with the left hand, grab the opponent's left lapel, and form an X with the wrists.

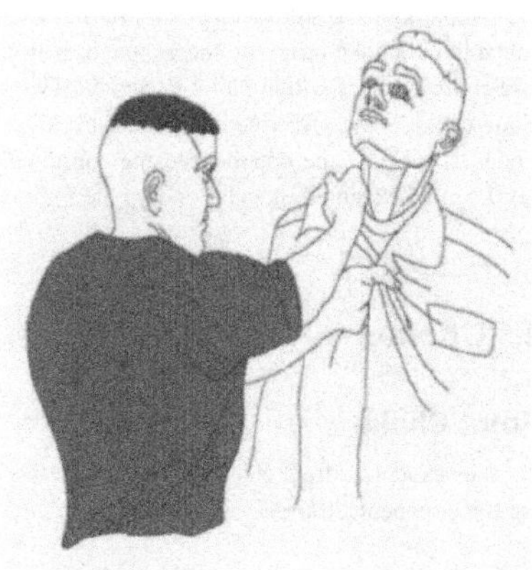

Press the right hand against the opponent's carotid artery.

Pull his lapel to the left with the left hand.

Side Choke

Marines execute a side choke when they are facing the opponent. The side choke is particularly effective when deflecting a punch thrown by an opponent. To execute the side choke, Marines—

Use the left hand to parry the opponent's arm inboard (to the inside of the opponent's reach).

Bring the right arm underneath the opponent's arm and up around the front of his neck.

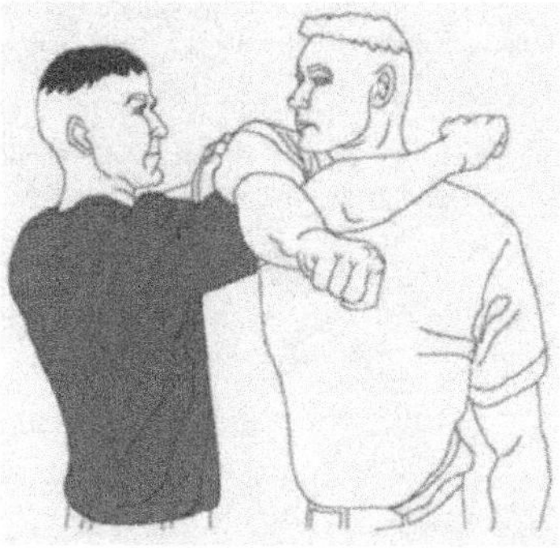

Extend the fingers, place the back of the forearm against the opponent's neck just below his ear, and press the carotid artery on his neck.

Note: If Marines are unable to place their fingers against the opponent's neck, they may use the back of the thumb or wrist.

Reach, with the left hand, around the back of the opponent's neck and clasp the hands together.

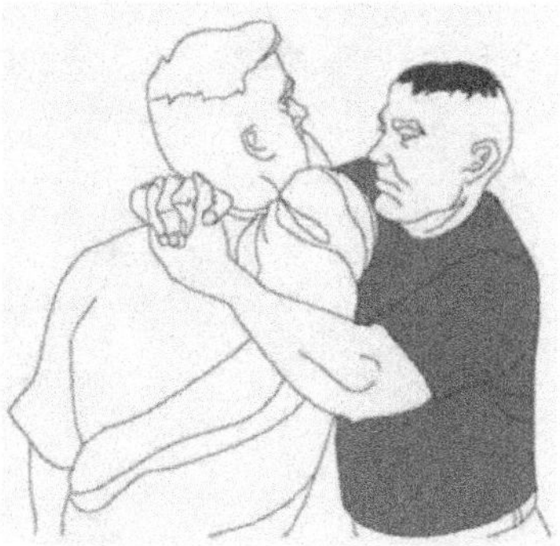

Pull the opponent toward the chest by pulling the clasped hands toward the chest.

Exert pressure on the side of the opponent's neck with the forearm.

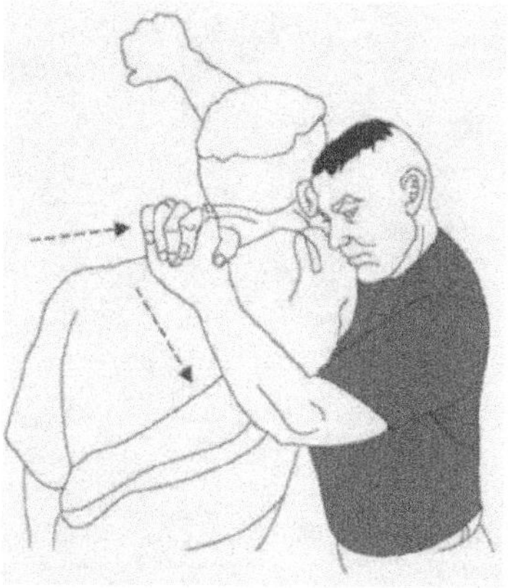

Rear Choke

Marines execute a rear choke when they are behind the opponent, the opponent is on the ground, or when they are taking the opponent to the ground. To execute the rear choke, Marines—

Reach, with the right arm, over the opponent's right shoulder and hook the bend of the arm around his neck.

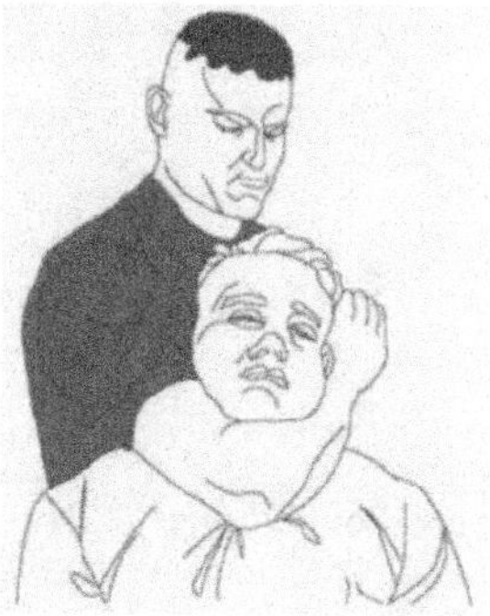

Clasp both hands together.

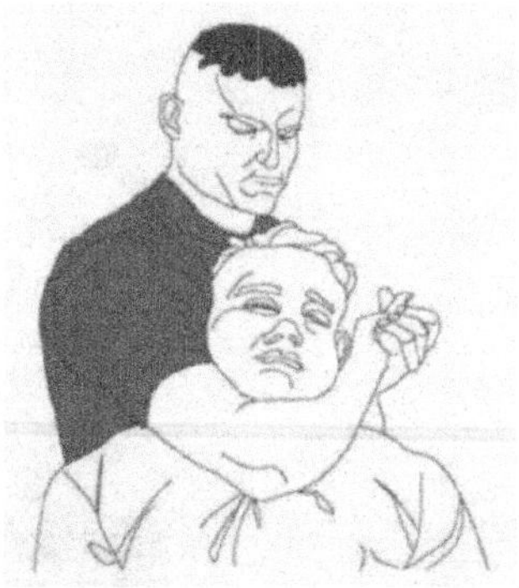

Exert pressure with the biceps and forearms on both sides of the opponent's neck on his carotid arteries.

Maintain pressure with the biceps and forearms on both sides of the neck and draw the opponent closer by drawing the right arm in.

Figure-Four Choke

A variation of the rear choke is the figure-four choke. The figure-four choke allows Marines to gain more leverage than the rear choke. If the rear choke cannot be secured, the figure-four variation is employed to increase the pressure of the choke on the opponent. To execute the figure-four variation of the rear choke, Marines—

Apply a rear choke. The Marine's body should be against the opponent's body.

Grasp the left biceps with the right hand and place the left hand against the back of the opponent's head.

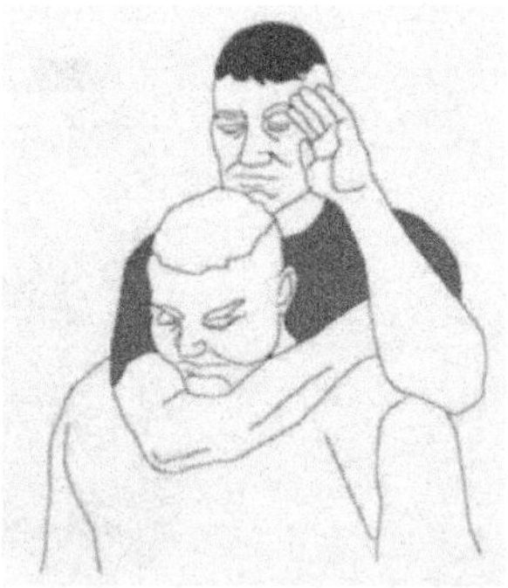

Push the opponent's head forward and down with the left hand.

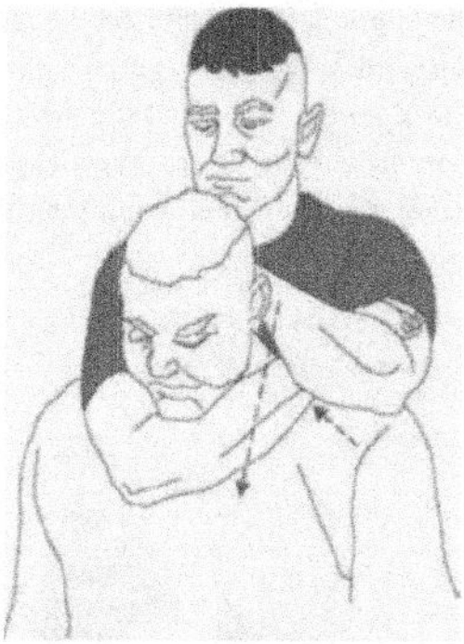

Draw the right arm in, maintaining pressure with the bicep and forearm on both sides of the opponent's neck.

3. Counters to Chokes and Holds

During a close combat situation, an opponent may apply a choke or hold on a Marine. If the opponent correctly applies a choke, a Marine quickly loses consciousness. If a choke is not executed properly, it often results in a hold, typically a bear hug or a headlock. A hold allows the opponent to control a Marine and removes the Marine's ability to attack. It is important for Marines to extract themselves from chokes and holds, regain the tactical advantage, and counter with strikes.

Although a choke causes unconsciousness in 8 to 13 seconds for a blood choke and 2 to 3 minutes for an air choke, the first movement in any counter to a choke is to clear the airway. Marines use softening techniques to loosen an opponent's grip and to clear their airway. Softening techniques are particularly effective if Marines lack the physical strength of their opponent. These techniques include groin strikes, eye gouges, foot stomps, etc. Softening techniques are not offensive; rather, they are used to loosen an opponent's hold.

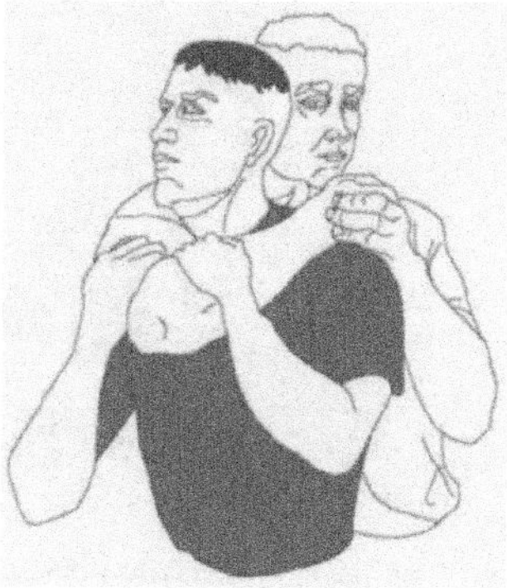

The second movement is to tuck the chin. Once the airway is clear, Marines tuck their chins to prevent the opponent from reapplying the choke.

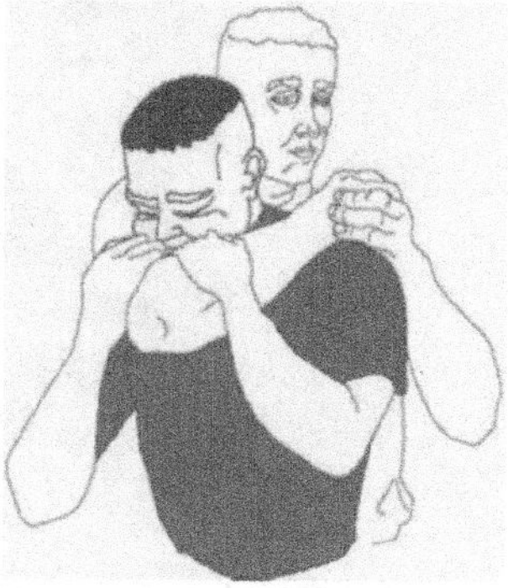

Counter to a Front Choke

Marines use a counter to a front choke when the opponent approaches from the front and uses both hands to choke a Marine around the throat. To execute the counter to the front choke, Marines—

Grasp the opponent's right forearm (where the elbow bends) with the left hand and apply downward pressure on the opponent's radial nerve with the fingers.

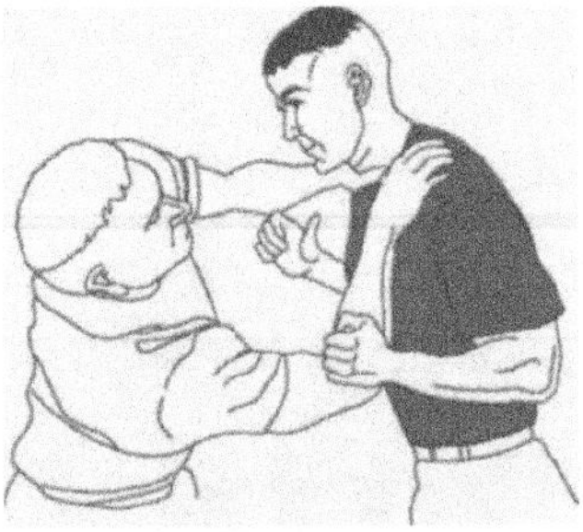

Execute a chin jab to the opponent's chin with the right hand. To generate power into the strike, bring the left foot to the outside of the opponent's right foot and rotate the hips into the strike.

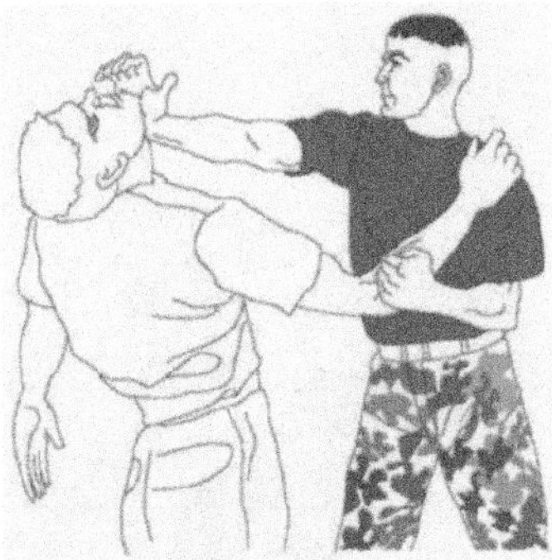

Counter to a Rear Choke

Marines execute a counter to a rear choke when the opponent approaches from the rear and puts his right arm around a Marine's throat. To execute the counter to the rear choke, Marines-

Grasp the opponent's forearm (at the radial nerve) and bicep with both hands and pull down just enough to clear the airway. Once the airway is clear, tuck the chin to protect the airway and to prevent the opponent from reapplying the choke.

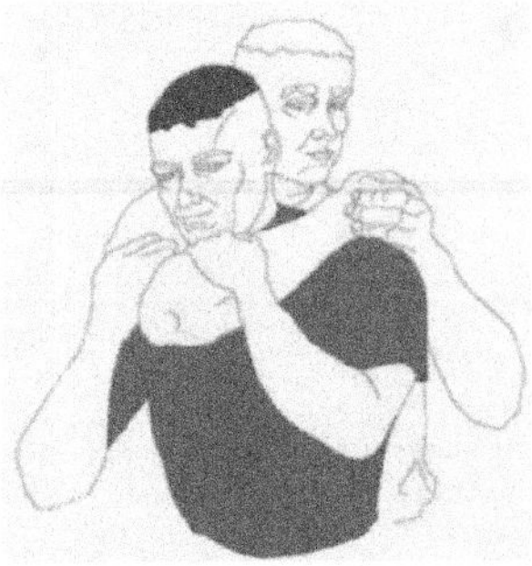

Step behind the opponent's right leg with the left foot, keeping both legs bent (almost in a squatting position).

Strike and drive the left elbow into the opponent's torso while rotating the hips and pivoting to the left, throwing the opponent back and to the ground.

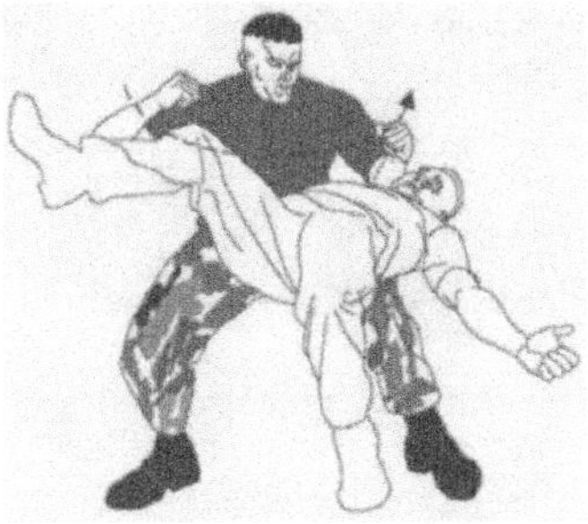

Counter to a Front Headlock

Marines use a counter to a front headlock when the opponent approaches from the front and puts his right arm around the Marine's neck, bends the Marine forward, and locks the Marine's head against his hip. To execute the counter to a front headlock, Marines—

Grasp the opponent's wrist and forearm with both hands and pull down to clear the airway. Maintain control of the opponent's wrist throughout the move. Once the airway is clear, tuck the chin to protect the airway and to prevent the opponent from reapplying the choke.

Move the right hand and arm across the opponent's torso.

Step forward and to the left with the left foot at a 45-degree angle.

Execute a sweep with the right foot against the opponent's right leg. At the same time, push against the opponent's chest with the right arm and shoulder to generate power in the sweep.

Counter to a Rear Headlock

Marines use a counter to a rear headlock when the opponent approaches from the rear and puts his right arm around the Marine's neck, bends the Marine forward, and locks the Marine's head against his hip. To execute the counter to the rear headlock, Marines-

Grasp the opponent's wrist and forearm with the right hand and pull down to clear the airway. Once the airway is clear, tuck the chin to protect the airway and to prevent the opponent from re-applying the choke.

Reach over the opponent's right shoulder with the left arm.

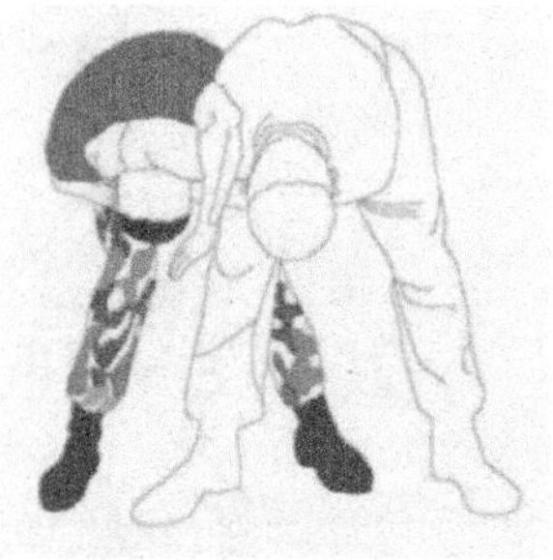

Grab any part of the opponent's face (chin, nose, eyes) and pull back while rising to a standing position.

Execute, with the right hand, a hammer fist strike to the opponent's exposed throat.

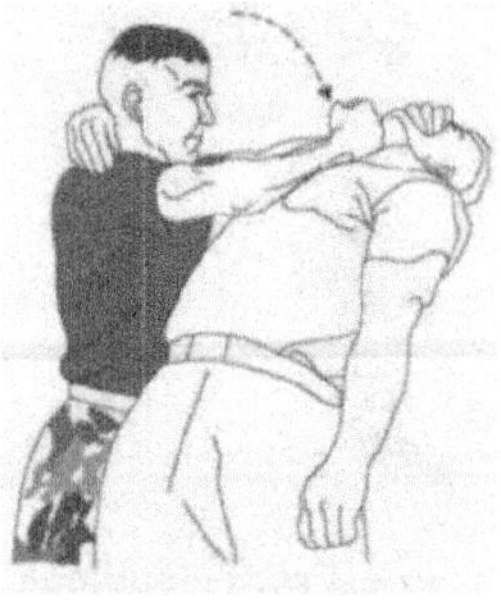

Counter to a Front Bear Hug

Marines execute a counter to a front bear hug when the opponent approaches from the front and puts both of his arms around the Marine's body, trapping the Marine's arms to the sides. To execute the counter to a front bear hug, Marines—

Step forward and to the left with the left foot at a 45-degree angle to the outside of the opponent's right leg, keeping the left leg bent.

Grasp the opponent's torso or arms to gain balance and to assist in throwing the opponent. It may be helpful to hook the opponent's right arm with the left arm.

Drive the right arm and shoulder forward and, at the same time, bring the right leg forward and sweep the opponent's right leg, bringing him to the ground.

Counter to a Rear Bear Hug

Marines use a counter to a rear bear hug when the opponent approaches from the rear and puts both of his arms around the Marine's body, trapping the Marine's arms to the sides. To execute the counter to a rear bear hug, Marines—

Step behind the opponent's right leg with the left foot, keeping both legs bent (almost in a squatting position). The left side of the body should be against the opponent's.

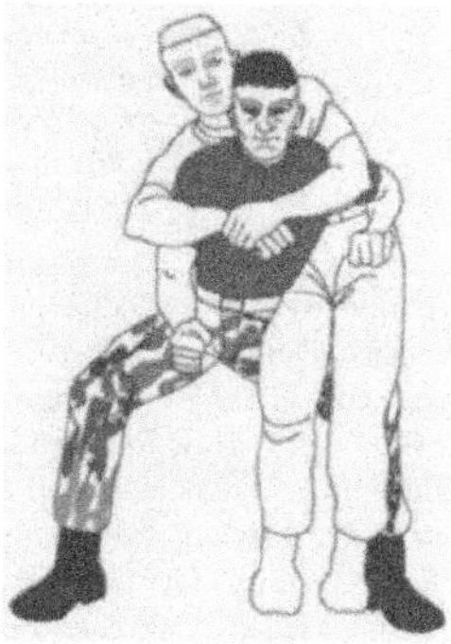

Pivot the hip, turning the body to the left and throwing the opponent back over the bent leg.

CHAPTER 7

GROUND FIGHTING

This chapter describes all techniques for a right-handed person. However, all techniques can be executed from either side.

In drawings, the Marine is depicted in woodland camouflage utilities; the opponent is depicted witlwut camouflage. In photographs, the Marine is depicted in woodland camouflage utilities; the opponent is depicted in desert camouflage utilities.

Marines should avoid being on the ground during a close combat situation because the battlefield may be covered with debris and there is an increased risk of injury. However, many close combat situations involve fighting on the ground. The priority in a ground fight is for Marines to get back on their feet as quickly as possible. In any ground fighting scenario, Marines will usually end up in one of four positions with the opponent. The offensive positions, in which Marines have a tactical advantage, are the guard and mount. The defensive positions, which are used as counters when the opponent has the tactical advantage, are the counter to the guard and the counter to the mount. Marines can also employ chokes during ground fighting to quickly end a fight.

1. Offensive Ground Fighting

Guard Position

Marines execute a guard position when the opponent is on top and the legs are wrapped around the opponent's legs. If the opponent is on top, he may try to choke the Marine, but the Marine still has the tactical advantage because the Marine is in a position to control the situation. To execute the guard position technique, Marines—

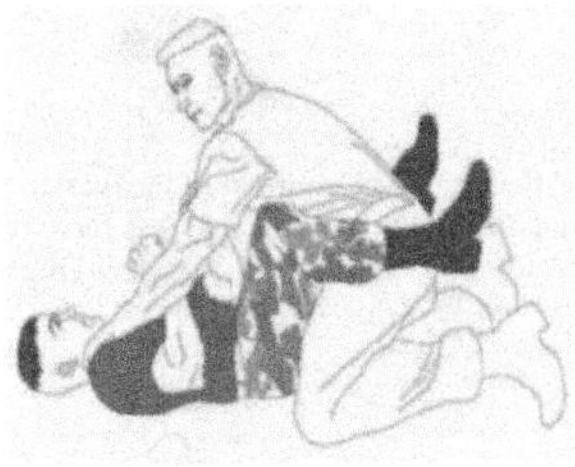

Trap the opponent's hands on the chest by crossing the hands on the chest so the forearms are resting on the opponent's forearms. Apply pressure with the elbows or forearms.

Strike the outside of the opponent's thigh with the cutting edge of the right heel. This causes the opponent to jerk to that side.

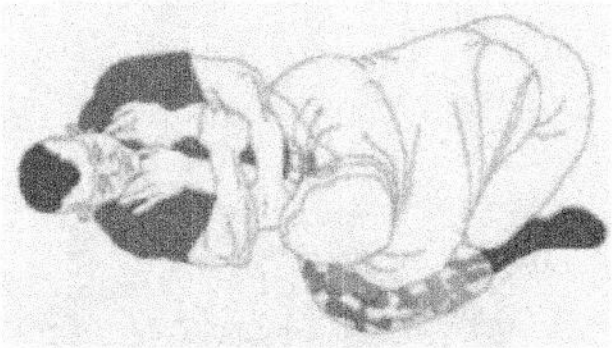

Move the head quickly to the left and swivel the hips to the right. At the same time, bring up both of the legs. Both legs are on the right side of the opponent's body.

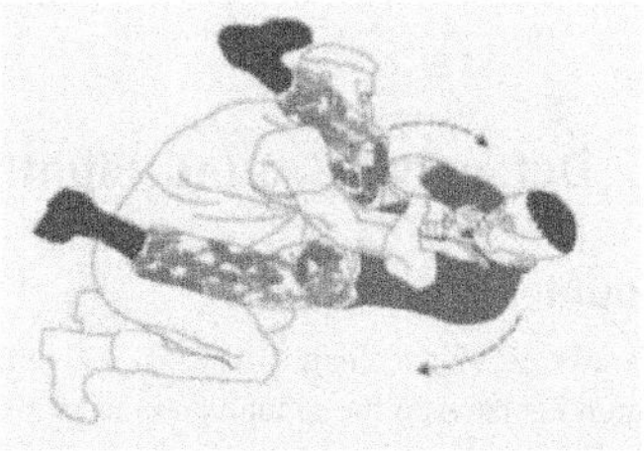

Bring the right leg down, hooking the opponent's neck and head. Exert downward pressure to roll him over on his back. Grasp and maintain control of the opponent's left arm. Upon completion of the move, the Marine is sitting up with legs bent over the opponent while maintaining control of his arm.

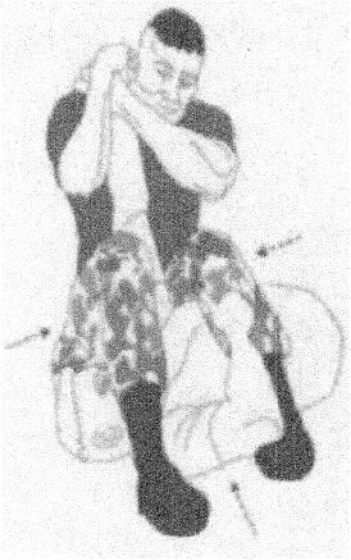

Keep legs and knees bent. Maintain pressure against the opponent's neck with the back of the right foot and against his side with the left foot underneath his armpit. Squeeze the knees together, locking the opponent' s arm.

Pull the opponent's arm straight up and fall back sharply, pulling his arm to the side in the direction of his little finger.

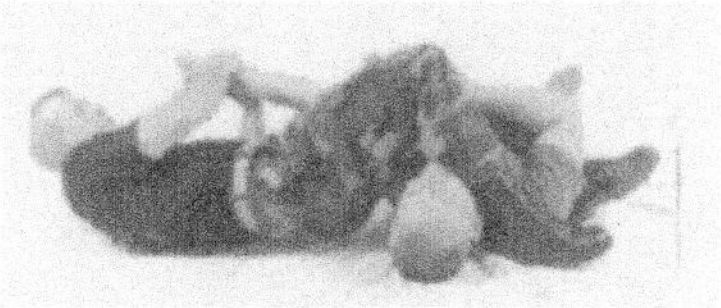

Return to standing.

Mount Position

Marines execute the mount if the opponent is lying on his back on the ground and the Marine is on top with legs wrapped around the opponent's body. This position is an offensive position because Marines are in a better position to control the opponent and to execute ground fighting techniques. To execute the mount position technique, Marines—

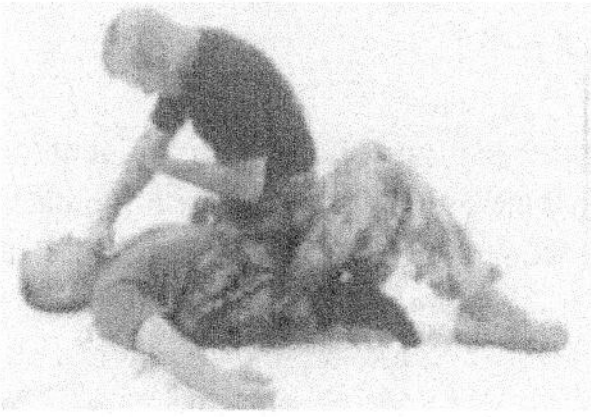

Grab the opponent's wrists or forearms with the hands. Hold them tightly against the torso.

Maintain control of the opponent's arms with the left arm and apply pressure to the opponent's brachial plexus (tie in) with the right hand.

Bring the left foot close to the opponent's right armpit and swing the right leg across the opponent' s head.

Keep the legs and knees bent. Maintain pressure against the opponent's neck with the back of the right foot and against his side with the left foot which is underneath his armpit. Squeeze the knees together, locking the opponent's arm.

Pull the opponent's arm straight up and fall back sharply, pulling his arm to the side in the direction of his little finger.

Return to standing.

2. Defensive Ground Fighting

Counter to the Guard

Marines use this technique if the opponent is lying on his back on the ground and the Marine is kneeling on the ground between the opponent's legs. To execute the counter to the guard position, Marines—

Strike the opponent's femoral nerve, located on the inside of the thigh, with the elbows. This forces the opponent to separate his legs.

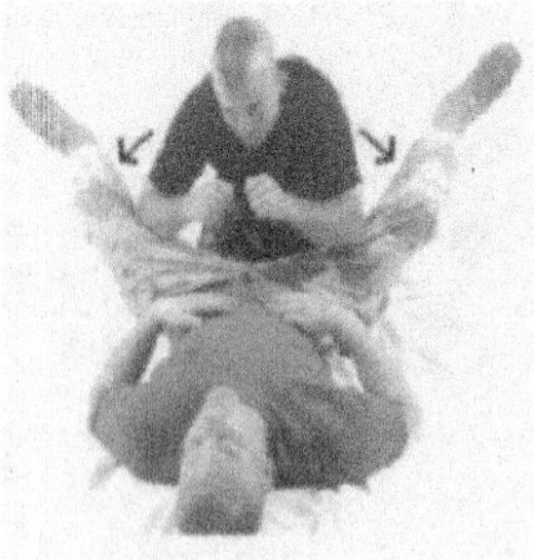

Strike the opponent's groin with the right fist.

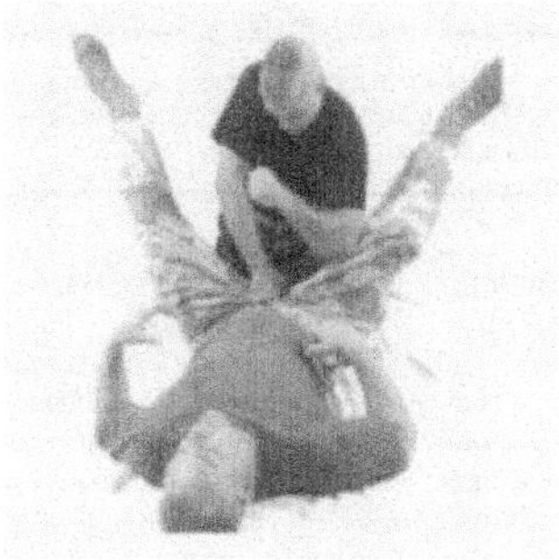

Hook the left arm underneath the opponent's right knee from the inside. Throw the opponent's leg over the head.

Duck the head quickly and move to the left.

Return to standing.

Counter to the Mount

Marines use this technique when they are lying on their back on the ground, the opponent is mounted on top of the Marine, and the opponent's legs are wrapped around the Marine. The opponent has the tactical advantage. To execute the counter to the mount position, Marines—

Grab the opponent's gear or clothing on his upper torso and pull him down close.

Use the right arm to hook the opponent's left arm, from the inside around the outside, above his elbow. With the right foot, hook the opponent's left leg or ankle.

Draw the elbow in to bend the opponent's elbow, bringing him down close. The arm must be hooked above the opponent's elbow in order to bend it.

Strike the opponent's side with the left hand.

Push the opponent over and roll him off to the right side.

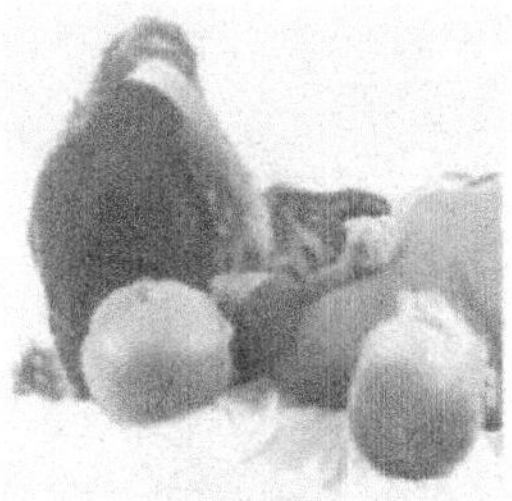

Maintain control of the opponent's hooked arm and move to a kneeling position.

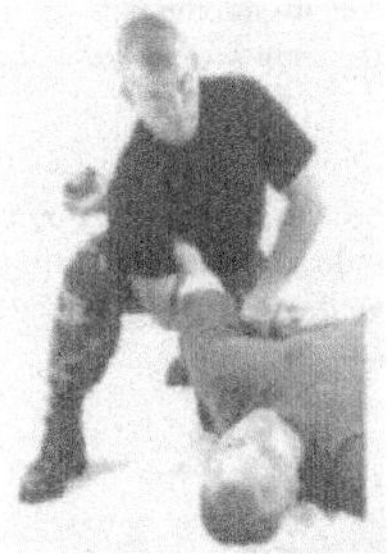

Move to a standing position while maintaining pressure on the opponent's arm and using the knee to apply pressure against the opponent's elbow.

3. Ground Fighting Chokes

WARNING

During training, never execute a choke at full force or full speed. Never hold a choke for more than 5 seconds.

The priority in ground fighting is for Marines to get back on their feet as quickly as possible. Sometimes, Marines can quickly end a ground fight by executing a choke on the opponent. When performed correctly, a choke can render an opponent unconscious in as little as 8 to 13 seconds. Chokes are easily performed regardless of size or gender. The chokes performed during ground fighting are the same as those performed while standing (see chap. 6).

Ground Fighting Front Choke

The ground fighting front choke is a blood choke performed most effectively from the mount position. The front choke employs the opponent's lapels or collar to execute the choke. To execute the ground fighting front choke, Marines—

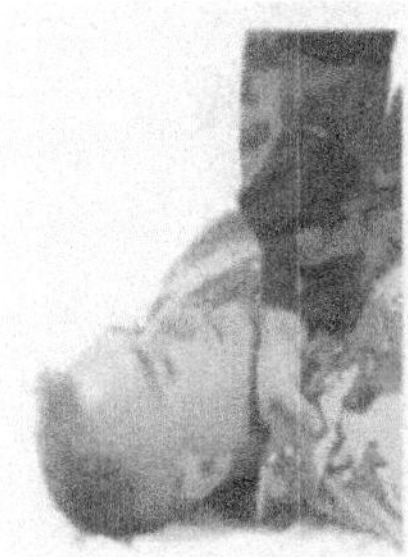

Grab the opponent's right lapel with the right hand, making certain that the knuckles or the back of the hand are against the carotid artery on the right side of the opponent's neck.

Keep the right hand pressed against the opponent's carotid artery, reach under the right arm with the left hand, and grab the opponent's left lapel, forming an X with the wrists.

Keep the right hand pressed against the opponent's carotid artery and pull the opponent's left lapel to the left with the left hand.

Ground Fighting Side Choke

The ground fighting side choke is a blood choke performed from the mount position. The ground fighting side choke is particularly effective when the opponent raises his arms and places them on the Marine's chest or throat. To execute the ground fighting side choke, Marines-

Use the left hand to parry the opponent's right arm inboard (to the inside of the opponent's reach). Bring the right arm underneath the opponent's arm and up around the front of his neck.

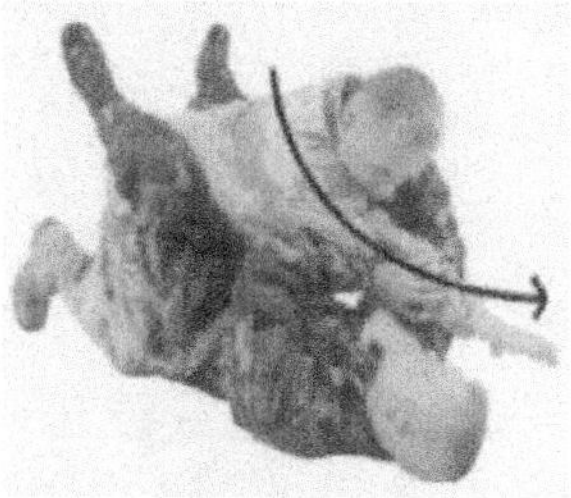

Keep the fingers extended, place the back of the forearm against the opponent's neck just below his ear and press the carotid artery on his neck. Depending on the Marine's or the opponent's body configuration, the back of the thumb or wrist may press against the opponent's neck.

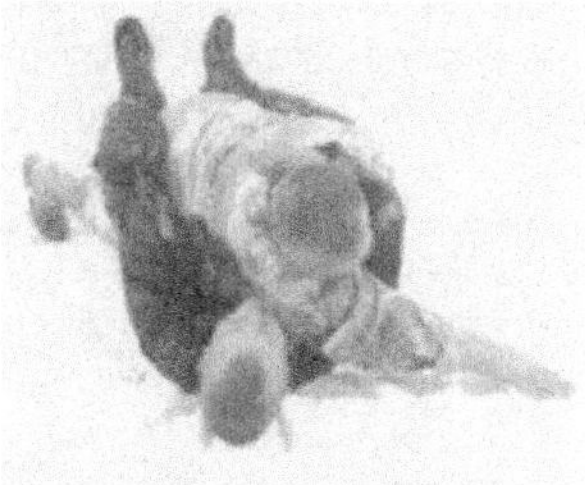

Use the left hand to reach around the back of the opponent's neck and clasp the hands together.

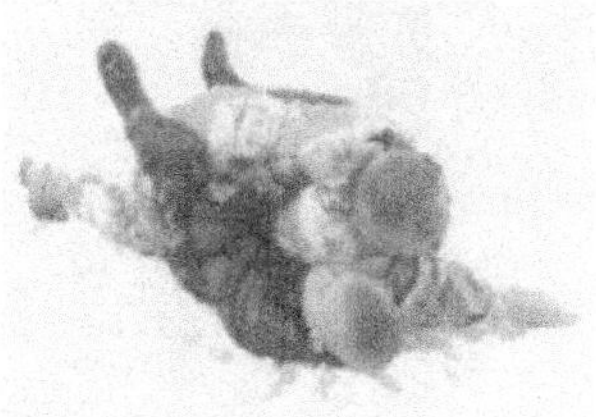

Pull the opponent toward the chest. Use the fore-arm to exert pressure on the side of his neck. This is done by pulling the clasped hands toward the chest.

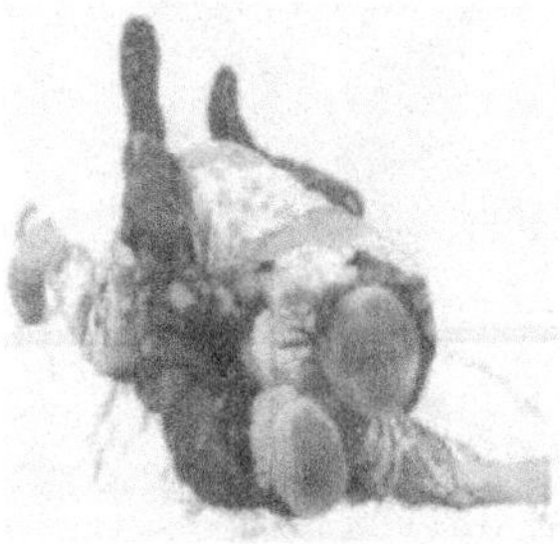

Ground Fighting Rear Choke

The ground fighting rear choke is a blood choke performed when Marines are behind the opponent. To execute the ground fighting rear choke, Marines—

Wrap the legs around the opponent with the left foot against the inside of his left thigh and the right foot against the inside of his right thigh.

Use the right arm to reach over the opponent's right shoulder and hook the bend of the arm around his neck.

Use the left hand to clasp both hands together.

Exert pressure with the biceps and forearms on both sides of the opponent's neck. This applies pressure to the opponent's carotid arteries. While maintaining pressure with the biceps and fore-arms, draw the opponent closer by drawing the right arm in. To increase the effectiveness of the choke, lean backward by arching the back and pulling the opponent back. At the same time, push the feet against the opponent's thighs.

Figure-Four Choke

A variation of the ground fighting rear choke is the figure-four choke. The figure-four choke allows Marines to gain more leverage on the rear choke. If the rear choke cannot be secured, the figure-four variation may be applied to increase the pressure of the choke on the opponent. To execute the figure-four variation of the rear choke, Marines—

Apply the ground fighting rear choke. The chest should be against the opponent's back.

Grasp the left biceps with the right hand and place the left hand against the back of the opponent's head.

Use the left hand to push the opponent's head forward and down.

Draw the right arm in while maintaining pressure with the biceps and forearm on both sides of the opponent's neck.

Increase the effectiveness of the choke by arching backward and pushing the feet against the opponent's thighs. Continue to pull the opponent back with the right arm while exerting pressure forward with the left hand against the opponent's head.

CHAPTER 8

NONLETHAL TECHNIQUES

This chapter describes all techniques for a riglit-handed person. However, all techniques can be executed from either side.

In drawings, the Marine is depicted in woodland camouflage utilities; the opponent is depicted without camouflage. In photographs, the Marine is depicted in woodland camouflage utilities; the opponent is depicted in desert camouflage utilities.

The Marine Corps' involvement in military operations other than war—e.g., humanitarian, peacekeeping, or evacuation missions—has greatly increased. These missions require skills that span the spectrum of conflict and support operations within a continuum of force. But the Corps' day-to-day existence also demands a responsible use of force. Nonlethal techniques are among the skills Marines use to apply a responsible use of force.

1. Unarmed Restraints and Manipulation

Marines operate within a continuum of force, particularly in support of peacekeeping or humanitarian types of missions. In these situations, Marines must act responsibly to handle a situation without resorting to deadly force. Unarmed restraints and manipulation techniques including joint manipulation, come-alongs, and takedowns can be used to control a subject without resorting to deadly force. Marines must train to become proficient in nonlethal techniques and to respond in a responsible manner. These techniques are referred to as compliance techniques, and they are applied in the third level in the continuum of force.

WARNING

During training, never apply the techniques for unarmed restraints and manipulation at full force or full speed. Use a slow and steady pressure to avoid injury.

Compliance Techniques

Compliance techniques are unarmed restraint and manipulation techniques used to physically force a subject or opponent to comply. Compliance can be achieved through the close combat techniques of-

Pain compliance using joint manipulation and pressure points. (Pain compliance is the initiation of pain to get compliance on the part of the subject.)
Come-along holds.

Principles of Joint Manipulation

Joint manipulation is used to initiate pain compliance and gain control of a subject. It involves the application of pressure on thejoints (elbow, wrist, shoulder, knee, ankle, and fingers). Pressure is applied in two ways:

In the direction in which the joint will not bend. For example, joints such as the knees

and elbows only bend in one direction and when pressure is applied in the opposite direction, pain compliance can be achieved. Beyond the point where the joint stops naturally in its range of movement (i.e., it no longer bends).

Since each joint has a breaking point, Marines should apply slow steady pressure only until pain compliance is reached. Continued pressure will break the joint and may escalate the violence of the situation.

Joint manipulation also uses the principle of off-balancing. A subject can be better controlled when he is knocked off balance.

Wristlocks

A wristlock is a joint manipulation that can be applied in a number of ways to achieve pain compliance. The wrist rotates in a number of directions and will bend in a single direction until its movement stops naturally. In a wristlock, pressure is exerted beyond that point by bending or twisting the joint. A wristlock is executed when an opponent tries to grab Marines or is successful in grabbing Marines or their equipment. A wristlock can also be performed by Marines if they wish to initiate control of an opponent.

Basic Wristlock. A basic wristlock is executed when Marines grab the opponent's left hand with the right hand. To execute the basic wristlock, Marines—

Use the right hand to grab the opponent's left hand by placing the thumb on the back of the opponent's hand so that the Marine's knuckles are facing to the left.

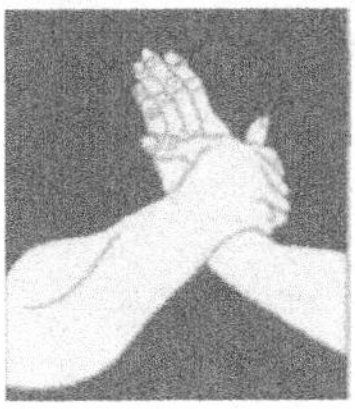

Hook the fingers across the fleshy part of the opponent's palm below the thumb. The fingers are used to anchor the hand so leverage can be applied to twist and bend the joint.

Exert dow nward pressure with the thumb to bend the opponcnt'sjoint. Rotate the opponent's hand to the right to twist the joint.

Use the left hand to further control the opponent.

Step in to the opponent to keep the opponent's hand in close to the body to control him and provide more leverage on the wristlock.

Note: When executing the basic wristlock with the left hand, the Marine grabs the opponent's hand so that the Marine's knuckles are facing to the right, and then rotates and twists the opponent's hand to thc left.

Reverse Wristlock. A reverse wristlock is executed when Marines grab the opponent's right hand with the right hand. To execute the reverse wristlock, Marines—

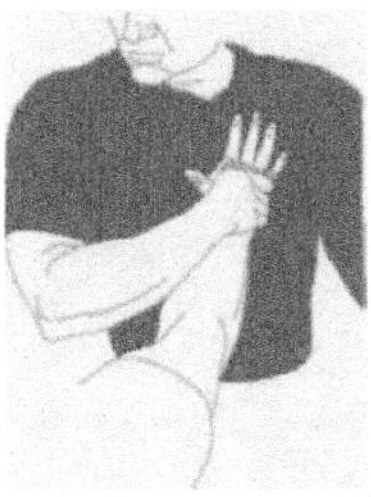

Place the right palm on the back of the opponent's right hand and wrap the fingers across the fleshy part of his palm below his little finger.

Twist the opponent's hand to the right while stepping in to place his hand against the chest. Apply downward pressure on the opponent's hand against the chest. Leave the opponent's hand on the chest to fully control the subject and to gain leverage.

Lean forward to use body weight to add additional pressure to the joint.

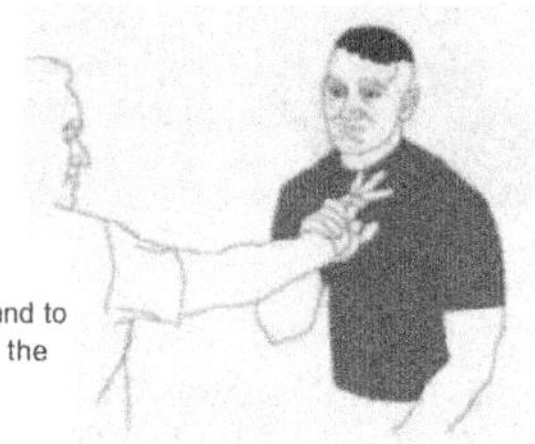

Use the left hand to further control the opponent.

Two-Handed Wristlock. Both hands can be used in the wristlock to maximize the leverage and pressure needed to bend and twist the joint. To execute the two-handed wristlock, Marines—

Place both thumbs on the back of the opponent's hand, thumbs crossed.

Hook the fingers of both hands around the fleshy part of the opponent's palm on both sides of his hand.

Step into the opponent and apply pressure downward on the back of his hand to bend the joint and rotate his wrist away from the body to twist the joint.

Enhanced Pain Compliance on Wristlock. Enhanced pain compliance techniques are applied in the third and fourth levels in the continuum of force. Additional pain be applied to a wristlock by-

Adding downward pressure to the elbow with the other hand or elbow by using the fingers to pull in on the opponent's radial nerve located on the inside of the forearm. When pressure is added to the opponent's radial nerve, his direction can be controlled.

Applying pressure against the opponent's finger joint to bend it in a direction it cannot bend (i.e., splitting the fingers).

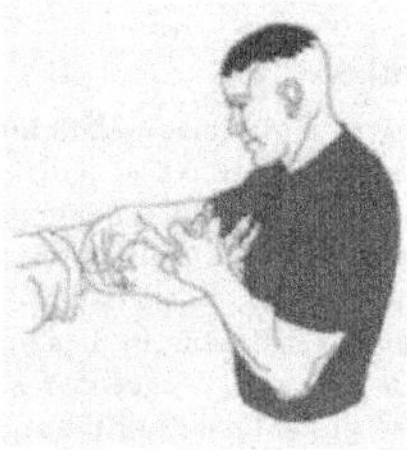

Come-Along Holds

Marines use a come-along hold to control and move an opponent.

Escort Position. A common come-along hold is the escort position. To execute the escort position, Marines—

Face the opponent. Use the left foot to step forward at a 45-degree angle. Turn to face the right side of the opponent.

Use the right hand to firmly grasp the opponent's right wrist. With the left hand, firmly grasp the opponent's right triceps.

Position the opponent's controlled arm diagonally across the torso, keeping his wrist against the right hip. The Marine should be standing to the right of and behind the opponent.

Note: This technique works well when escorting an opponent on either the right or left side. Take caution when escorting an opponent by ensuring his controlled hand is not in a position to grab the holstered weapon. The preferred escort position is from the Marine's left side, so that the opponent is kept further away from the weapon.

Wristlock Come-Along. To execute the wristlock come-along, Marines-

Use the left hand to execute a basic wristlock. Incorporate the right hand in a two-handed wristlock for more control.

Maintain pressure on the opponent's wrist with the right hand, step forward, and pivot around to stand next to the opponent.

Release the left hand, quickly reach under the opponent's arm from behind, and grab his hand.

Use the left hand and apply downward pressure on the opponent's wrist.

Controlling Technique. The following controlling technique is used when an opponent grabs the Marine's wrist. To execute the technique, Marines—

Trap the opponent's hand with the palm of the other hand.

Rotate the opponent's trapped hand up and on his forearm while maintaining downward pressure on his trapped hand.

Apply downward pressure with both hands until the opponent is taken to the ground.

Armbars

An armbar is a joint manipulation in which pressure is applied on a locked elbow, just above the joint, in the direction the joint will not bend. An armbar has to be locked in quickly, but still requires a slow, steady pressure to gain compliance.

Basic Armbar. To execute a basic armbar, Marines-

Use the right hand to grab the opponent's right wrist.

Bring the left hand down on or above the opponent's elbow joint. To gain additional leverage, pivot to face the opponent.

Use the left hand to apply downward pressure on or above the opponent's elbow joint while pulling up on his wrist.

Armbar from a Wristlock. To execute an armbar from a wristlock, Marines-

Use the right hand to grab the opponent's right hand and execute a reverse wristlock.

Bring the left hand down on or above the opponent's elbow joint.

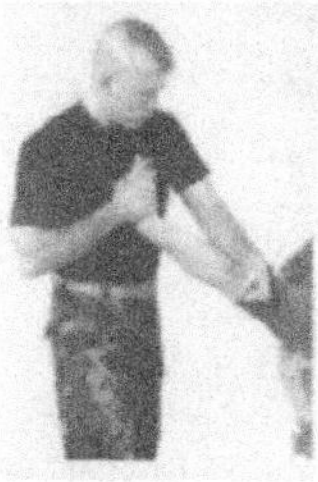

Use the left hand to apply downward pressure on or above the opponent's elbow joint while pulling up on his wrist.

Takedowns

A takedown is used to bring an opponent to the ground to further control him.

Takedown From a Wristlock Come-Along. To take the opponent to the ground from a wristlock come-along, Marines—

Use the right foot to push down on the opponent's calf or Achilles tendon.

Maintain control of the opponent's wrist and elbow and apply a slow, steady pressure to bring him to the ground.

Armbar to a Takedown. This technique is used to take a noncompliant opponent to the ground from an armbar. To execute the armbar takedown, Marines—

Use the right hand to execute a reverse wristlock.

Bring the left hand or forearm down on or above the opponent's elbow joint.

Pivot so the back is facing the opponent and, at the same time, lift the left elbow and slide the body so it is against the opponent, placing the armpit high above the opponent's elbow joint.

Lean back, placing the body weight on the opponent's arm until he complies or is taken to the ground.

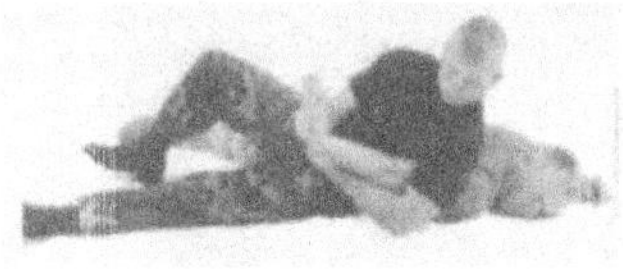

Note: This technique may break the opponent's arm. Therefore, this technique should not be employed if the objective is a nonlethal takedown.

Wristlock Takedown. This technique is used to take a noncompliant opponent to the ground from a basic wristlock and to put him in a position where he can be handcuffed, if necessary. To execute the takedown, Marines-

Use the right hand to execute a basic wristlock.

Incorporate the left hand in a two-handed wristlock.

Apply downward pressure on the wristlock, pivot on the ball of the right foot, and quickly turn to the right to take the opponent to the ground.

Continue to apply pressure on the wristjoint as the opponent lands on his back with his arm straight in the air.

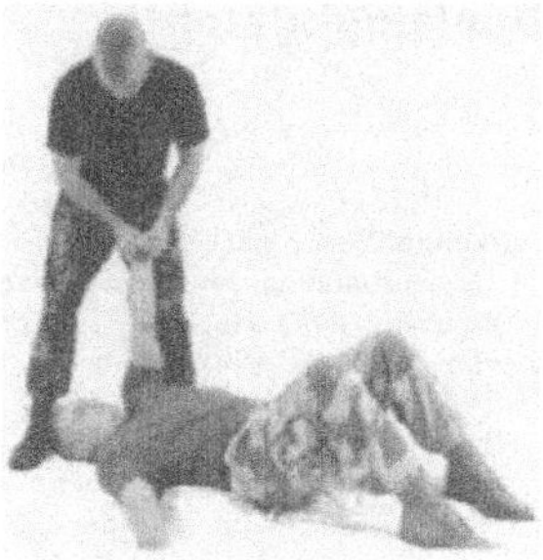

Slide the left foot under the opponent's back, underneath his armpit.

Apply pressure with the knee against the opponent's triceps while pulling back on his arm.

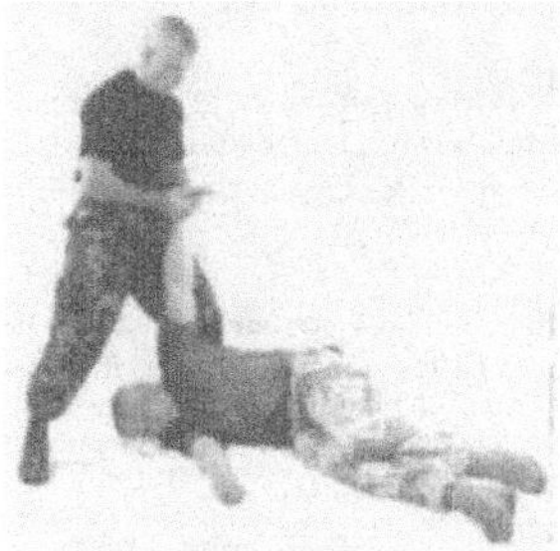

Continue applying pressure with the knee on the opponent's arm. Pivot and step around the opponent's arm to roll him on his stomach.

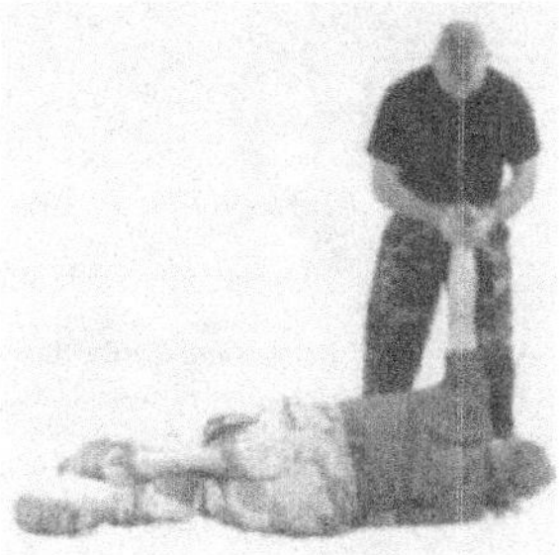

Kneel down with one knee on the opponent's back. The other knee is placed on the opponent's neck and shoulder on either side of his arm. Apply inward pressure with the knees to lock his arm in place.

Tell the opponent to put his other hand in the middle of his back. Bring the opponent's controlled hand to the center of his back.

Escort Position Takedown. This technique can be used to control a noncompliant opponent from the escort position. To execute the escort position takedown, Marines—

Lock the opponent's arm straight across the body while rotating his wrist away from the body.

Use the left hand or forearm and apply downward pressure above the opponent's elbow where the triceps meet.

Step back with the right foot and, keeping the opponent's hand controlled against the hip, pivot to the right while continuing to apply downward pressure on his arm to bring him to the ground.

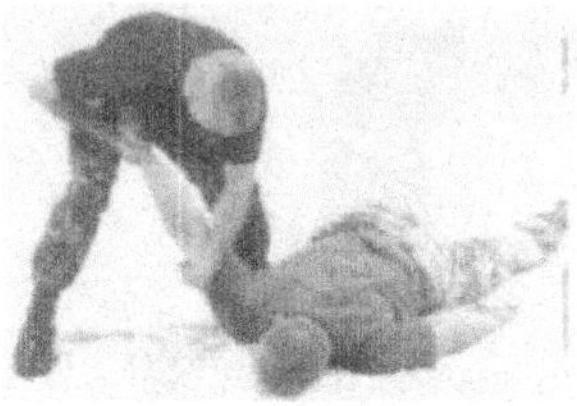

Note: This technique works well when escorting a subject on the Marine's right or left side. When taking down a subject from the right side, step back and pivot to the left.

2. Nonlethal Baton

A baton or nightstick can be an effective compliance tool when used correctly. Batons can be used defensively (blocking), offensively (striking), and as a restraining device when needed. In the fourth level of the continuum of force (assaultive [bodily harm]), defensive tactics include baton or nightstick blocks and blows. Blows to the head or other bony parts of the body are considered deadly force. When deadly force is not authorized, Marines must be able to employ blocks, strikes, and restraints with the baton with the minimum of force.

Grips

One-Handed Grip. To execute the one-handed grip, Marines—

Use the right hand to grasp the lower end of the baton, about 2 inches from the end.

Wrap the thumb and index finger around the baton so they are touching one another. The grip on the baton should be firm, but natural.

Two-Handed Grip. To execute the two-handed grip, Marines-

Use the right hand to grasp the lower end of the baton, about 2 inches from the end. Wrap the thumb and index finger around the baton so they touch one another.

Use the left hand to grasp the upper end of the baton, palm down, about 2 inches from the end. The hands should be approximately 10 to 12 inches apart.

Stance and Method of Carry

The basic warrior stance serves as the foundation for initiating nonlethal baton techniques. The method of cany provides effective defensive positions with a wide range of options to control a combative opponent.

One-Handed Carry. To execute the one-handed carry, Marines—

Grip the baton using the one-handed grip.

Elevate the baton, with the gripping hand at a level between the belt and shoulder.

Keep the left hand in the position of the basic warrior stance.

Two-Handed Carry. This carry is effective for blocks. To execute the two-handed carry, Marines-

Grip the baton using the two-handed grip.

Elevate the baton, with the left hand higher than the right hand.

Orient the weapon toward the subject.

Movement

In a nonlethal confrontation, movement may be made to create distance between Marines and the opponent or to close the gap to control him. When facing an opponent, Marines move in a 45-degree angle to either side of the opponent. Moving at a 45-degree angle is the best way to both avoid an opponent's strike and to put Marines in the best position to control the opponent.

Blocking Techniques

One-handed blocks are used when carrying the baton in the one-handed carry. The same one-handed blocks used in combative stick techniques (see page 3-6) apply in nonlethal baton engagements. The blocks include blocks for a vertical strike, a forward strike, and a reverse strike. Because the baton is often carried with two hands, there are also two-handed blocks that are effectively used from this carry. Two-handed blocks are discussed in the following subparagraphs.

High Block. Marines execute a high block to deter a downward vertical attack directed at the head and shoulders. To execute the high block, Marines—

Raise the baton up to a level even with or above the head. The baton should be in a horizontal position to block the blow. The fingers of the left hand should be open and behind the baton.

Place the baton perpendicular to the opponent's striking surface to absorb the impact of the blow.

Bend the elbows slightly to help absorb the impact of the blow. The arms should give with the strike of the blow.

Low Block. Marines execute a low block to deter an upward vertical attack directed at the abdomen, groin, or torso. The opponent's blow can be delivered by a foot, knee, or fist. To execute the low block, Marines—

Lower the baton to a level even with or below the groin. The baton should be in a horizontal position to block the blow. The fingers of the left hand should be open and behind the baton.

Place the baton perpendicular to the opponent's striking surface to absorb the impact of the blow.

Bend the elbows slightly to help absorb the impact of the blow. The arms should give with the strike of the blow.

Right Block. Marines execute a right block to deter an attack directed at the head, neck, flank, or hip. The opponent's blow can be delivered by a foot, knee, fist, or elbow. To execute the right block, Marines-

Thrust the baton in a vertical position to the right side.

Pivot to the right by stepping with the left foot and pivoting off the ball of the right foot. Rotate the hips and shoulders into the direction of the block. The fingers of the left hand should be open and behind the baton.

Place the baton perpendicular to the opponent's striking surface to absorb the impact of the blow.

Bend the elbows slightly to help absorb the impact of the blow. The arms should give with the strike of the blow.

Left Block. Marines execute a left block to deter an attack directed at the head, neck, flank, or hip. The opponent's blow can be delivered by a foot, knee, fist, or elbow. To execute the left block, Marines—

Thrust the baton in a vertical position to the left side.

Pivot to the left by stepping with the right foot and pivoting off the ball of the left foot. Rotate the hips and shoulders into the direction of the block. The fingers of the left hand should be open and behind the baton.

Place the baton perpendicular to the opponent's striking surface to absorb the impact of the blow.

Bend the elbows slightly to help absorb the impact of the blow. The arms should give with the strike of the blow.

Middle Block. Marines execute either a left or right block to deter an attack directed at the face, throat, chest, or abdomen. To execute the middle block, Marines—

Thrust the baton in a vertical position straight out in front of the body. The fingers of the left hand should be open and behind the baton.

Place the baton perpendicular to the opponent's striking surface to absorb the impact of the blow. The baton should be held with the left hand forward of the right.

Bend the elbows slightly to help absorb the impact of the blow. The arms should give with the strike of the blow.

Restraining Technique

The strong-side armlock is used to restrain an opponent who is not compliant. To execute the strong-side armlock, Marines—

Use the right hand to run the baton up under the opponent's left armpit, parallel to the ground.

Use the right foot to step forward at a 45-degree angle to the left side of the opponent. The baton is across his forearm.

Use the right hand to drive the baton forward and up so the action bends the opponent's arm behind his back. At the same time, continue moving around him to get behind him.

Place the baton on the opponent's forearm with the thumb and/or fingers and apply pressure to his forearm. At the same time, grasp the other end of the baton with the left hand.

Pull up on the low end of the baton with the right hand. At the same time, push down on the top end of the baton with the left forearm, reaching around with the left hand to grasp the opponent's biceps or shoulder.

Continue exerting downward pressure with the left forearm while pulling back on the opponent's biceps with the left hand. This places the opponent in a position where he is controlled and can be moved.

Apply pressure with the foot against the bend in the opponent's leg above his calf. This lowers the opponent to the ground rather than throwing him to the ground and risking severe injury.

Striking Target Areas

Marines must avoid striking an opponent in the head, neck, or other bony parts with the baton because this is considered deadly force and can lead to serious bodily injury or death. Instead, the legs, arms, and buttocks are target areas that are considered nonlethal.

Legs. Primary targets are the thighs and lower legs. Avoid striking the knee and ankle joints because this can cause permanent damage.

Arms. Primary targets are the upper arms. Avoid striking the shoulder and elbow and wrist joints because this can cause permanent damage.

Buttocks. Primary targets are the buttocks. Avoid striking any other part of the torso, including the chest, rib cage, spine, tail bone, and groin because strikes to these areas can cause permanent damage or death.

One-Handed Striking Techniques

One-Handed Forward Strike. A forward strike follows either a horizontal line or a downward diagonal line using a forehand stroke. To execute the one-handed forward strike, Marines-

Stand facing the opponent with the baton carried in a one-handed cany.

Place the right hand palm up, swing the baton from right to left, and make contact with the opponent.

One-Handed Reverse Strike. A reverse strike follows either a horizontal line or a downward diagonal line using a backhand stroke. To execute the one-handed reverse strike, Marines—

Stand facing the opponent with the baton carried in a one-handed carry.

Bend the right arm and cross the arm to the left side of the body. The baton should be close to or over the left shoulder.

Place the right hand palm down, swing the baton from left to right, and make contact with the opponent.

Two-Handed Striking Techniques

Two-Handed Forward Strike. This strike is an effective follow-up to a middle block or left block. To execute the two-handed forward strike, Marines—

Pull back with the left hand while driving the right hand forward toward the opponent. The baton should be horizontal to the ground. Power is generated by stepping forward with the right foot and rotating the right hip and shoulder into the strike.

Contact the opponent with the end of the baton.

Two-Handed Reverse Strike. This strike is an effective follow-up to a middle block or left block. To execute the two-handed reverse strike, Marines—

Pull back with the right hand while driving the left hand forward toward the opponent. The baton should be horizontal to the ground. Power is generated by stepping forward slightly with the left foot and rotating the left hip and shoulder into the strike.

Contact the subject with the end of the baton.

Front Jab. This strike is effective for countering a frontal attack. It can also be executed as a quick poke to keep a subject away. To execute the front jab, Marines thrust both hands forward in a quick jab. The baton is held either horizontal to the ground or at a slight downward angle.

Rear Jab. This strike is effective for countering a bear hug from the rear. To execute the rear jab, Marines thrust both hands rearward in a quick jab. The baton is held either horizontal to the ground or at a slight downward angle.

APPENDIX A

PUGIL STICK TRAINING

A pugil stick is a training device used to simulate a rifle bayonet so that effective, but safe, training can be conducted to build proficiency of rifle bayonet techniques. Pugil stick training builds on the techniques used to throw punches. Pugil stick training is the only "full contact" training provided to Marines in the Close Combat Program. Pugil stick training teaches Marines to function when faced with stress and violence, and it prepares them to deliver a blow and take a blow. It also provides them with the physical and mental skills vital to success on the battlefield.

1. Pugil Stick Training

Design

A pugil stick consists of a stick wrapped in padding at both ends that can be gripped like a rifle. The pugil stick is approximately the same weight and length of an unloaded rifle with a bayonet attached.

Holding the Pugil Stick

The pugil stick is held in the same manner as the sen ice rifle. All movements come from the basic warrior stance. To hold the pugil stick correctly, Marines—

Use the right hand to grasp the lower end of the pugil stick overhanded.
Use the left hand to grasp the upper end of the pugil stick underhanded.
Use the right forearm to lock the lower end of the pugil stick against the hip.
Orient the blade end of the pugil stick toward the opponent.

Safety Equipment

The following safety equipment must be worn during any pugil stick bout.

Groin Protection. Groin protection protects the groin from an accidental blow. It should be pulled high around the waist to protect the groin area, with the concave portion against the body.

Flak Jacket. The flak jacket provides protection to the body. It is worn completely fastened.

Neck Roll. The neck roll prevents whiplash if Marines receive a blow to the head. The neck roll further supports the head and protects the neck from blows. The neck roll is worn above the flak jacket and below the helmet. The tied end of the neck roll faces the front.

Helmet. A regulation football helmet protects the face and head. The helmet must fit snugly and the chin straps must be adjusted and snapped.

Mouthpiece. The mouthpiece is worn on the upper teeth to protect the teeth.

Gloves. Marines wear gloves to protect the hands if pugil sticks do not have gloves built-in.

Pugil Stick Screening

Prior to pugil stick training, instructors must ask participating Marines the following questions. If a Marine answers "yes" to any question, he must see the corpsman or a competent medical authority who evaluates him and determines whether or not the he can participate in the training.

- Are you on light duty?
- Are you restricted to running shoes by a corpsman or a doctor?
- Have you fought in a pugil stick bout within the last 7 days?
- Have you received a blow to the head within the last 7 days?
- Have you had a concussion within the last 6 months?
- Have you had dental surgery within the last 24 hours?
- Do you have stitches or staples on your body?
- Have you had a shoulder or head injury within the last 5 years?
- Are you taking a prescription drug?
- Do you have an ear infection or current sinus infection?
- Have you had a broken bone within the last 6 months?

Safety Personnel

The following safety personnel are required to conduct pugil stick training:

- One close combat instructor must officiate the bout.
- One close combat instructor trainer, commissioned officer, or staff noncommissioned officer must be in the training area to serve as range safety officer.
- One corpsman must be in the training area.

Training Area

To prevent injury, Marines train on areas with soft footing (i.e., sand or grass). Training mats are not recommended because feet can stick to the mats, prohibiting movement or causing joint injuries by twisting a knee. Bouts should not take place on a hard surface area; e.g., a flight deck or parking lot. A boxing ring may be used to conduct pugil stick bouts; ring dimensions can vary as long as there is ample room to execute the techniques in the training area.

Second Impact Syndrome

Second Impact Syndrome occurs when a second blow to the head produces a second concussion that occurs within 1 week following a previous concussion (before recovery from the first concussion). Second Impact Syndrome causes rapid brain swelling and can cause death. Therefore, there must be 7 days between pugil stick bouts to reduce the risk of severe injury resulting from Second Impact Syndrome. The 7 day separation between pugil stick bouts significantly reduces the possibility of injury, particularly in someone who may have suffered a brain injury or concussion but shows no symptoms.

Any Marine who experiences headaches or the following symptoms after training must be examined by appropriate medical personnel:

- Blurred vision.
- Ringing in the ears.
- Dilation of the pupils.
- Slurred speech.
- Bleeding from ears or mouth.
- Swelling in head or neck area.
- Any unnatural discoloration in head or neck area.

The Marine should not be allowed to participate in pugil stick training or any other activity where a heavy blow might be sustained for a minimum of 7 days after the headache or other symptoms have subsided.

Safety Measures

The following safety measures must be followed:

- Gear must be worn properly throughout training.
- Contact lenses or glasses will not be worn.
- False teeth will be removed from the mouth.
- Nothing will be worn around the neck except the neck roll.

Competition among groups of Marines is authorized as long as it does not overshadow training objectives or compromise safety procedures.

Safety and proper techniques are paramount.

Safety is more important than competition.

2. General Rules and Regulations Governing Pugil Stick Bouts

Instructors and Support Personnel Requirements

At a minimum, one close combat instructor observes a pugil stick bout. For safety purposes, it is better to have two instructors judging a bout because each instructor can fully observe each of the fighters. The best position for observation is to the right of a fighter. This allows the instructor to see the fighter's facial expression and body movement. The instructor's position must not interfere with the fight.

In addition, one commissioned or staff noncommissioned officer (to serve as range safety officer) and one corpsman will be in the training area.

Prior to the Bout

Prior to the bout, Marines—

Are paired according to height, weight, and gender.

Wear the proper safety equipment.

Once Marines have properly donned the safety equipment, they wait for instructions from the close combat instructor. On the command of the close combat instructor, a pair of Marines will enter the ring. Once in the ring, the close combat instructor inspects each Marine for the proper safety equipment.

Note: There should be enough gear so that when two Marines are training, two other Marines can be donning safety equipment for the next bout.

During the Bout

The bout begins when the close combat instructor blows the whistle. All strikes are directed above the waist. Upon hearing a whistle blast, all fighting immediately ceases.

Stopping a Bout

There are two reasons for stopping a bout: delivery of a scoring blow or an unsafe condition. The close combat instructor trainer, close combat instructor, or the range safety officer may stop the bout at any time an unsafe condition is observed.

Scoring Blow. A scoring blow is an offensive technique delivered to a vulnerable area of an opponent with sufficient force and precision to be considered as a disabling or killing blow. Scoring blows are not judged solely on the degree of force with which a blow is delivered, but on the accuracy and techniques employed. A scoring blow is defined as—

A straight thrust with the blade end of the weapon (red end of the pugil stick) to the opponent's face mask or throat.

A slash to the side of the opponent's helmet (below the ear) or neck with the red end of the pugil stick.

A heavy blow to the opponent's head with an authorized technique (i.e., buttstroke, smash) using the butt (black end) of the pugil stick.

When a scoring blow is delivered, the close combat instructor blows the whistle to stop the bout.

Unsafe Condition. The bout will be stopped as soon as an unsafe condition exists. An unsafe condition exists when a Marine is unable to defend himself, loses his balance and falls down on one or two knees or falls down completely, shows instability (e.g., buckling at the knees), loses muscular tension in his neck and his head snaps back or to one side, or appears disoriented. An unsafe condition also exists if a Marine lets go of one end of the pugil stick, the equipment (e.g., helmet, neck roll) falls off, or a Marine fails to use the proper techniques.

If any of these conditions occur—

- The close combat instructor trainer, close combat instructor, range safety officer, or anyone supervising training stops the bout and separates the two Marines.
- The corpsman evaluates the possibly injured Marine to determine if the fight can continue. The corpsman—
  - Ensures that he is alert and responsive, both verbally and physically.
  - Talks to him to see if he is coherent.
  - Makes certain he comprehends and replies to verbal questions.
  - Makes certain his speech is not slurred.
  - Checks his physical signs.
  - Ensures his eyes are focused and not dazed or glazed.
  - Ensures his legs are not wobbly or shaky.

The pugil stick is not used as a baseball bat. The use of unauthorized techniques will result in expulsion from the bout. Only techniques taught in bayonet training are authorized. These techniques include-

- Straight thrust.
- Buttstroke (horizontal and vertical).
- Smash.
- Slash.
- Parry.
- Blocks (low, high, left, right).

After the Bout

A third whistle blast is used prior to identifying the Marine who delivered the killing blow.

A fourth whistle blast is used prior to demonstrating the killing technique used by the Marine who won the bout.

3. Directions for Making Pugil Sticks

Close combat instructor trainers are taught to make pugil sticks to standard specifications at school. It is recommended that all pugil sticks be made under the supervision of a close combat instructor trainer. Until the Marine Corps acquires a standardized pugil stick that units can order through the supply system, instructors use the following information and instructions to make pugil sticks locally.

Supplies

The following supplies are needed to make one pugil stick:

- One 45" by 2" circular, oak, wood dowel.
- Four 9" wide by 27" long foam pad pieces to make two inner handguards of double thickness.
- Four 11" wide by 19" long foam pad pieces to make two outer handguards of double thickness.
- Canvas material to cover inner and outer handguards.
- One foam pad cut 1/4" thick by 7" wide by 13" long to make the centerguard.
- One 8" wide by 14"long piece of canvas material to make a pugil stick sock.
- Canvas material to make two pugil stick end caps. Ideally, two different color fabrics should be used to simulate the bayonet and buttstock ends of a rifle. Common colors for end caps are black and red. Each end cap is a different color to help the officiating certified close combat instructor determine the scoring blow during the pugil stick bout.
- Two 1/4" thick by 11" wide by 54" long foam pad pieces to make two ends.
- Two foam pads cut 1/4" thick by 3" wide by 14 1/2" long to make two end plug inserts.
- One can of 3M Photo MountQ spray adhesive (NSN 8040-01-120-4009)to adhere foam padding to the stick. (Spray adhesive is available from self-service.)
- One roll of duct tape or riggers tape (NSN 7510-00-266-5016) to protect foam pad and

canvas from wear and to reinforce gluing action of spray adhesive. (Duct tape or riggers tape is available from self-service.)

One roll para-cord or 5-50 cord to tie both the inner and outer handguards together.

One roll of thread strong enough to sew canvas material together.

Sixteen grommets, one for each of the four corners of the inner and outer handguards.

One foam sleeping mat (NSN 8465-01-109-3369). Use unserviceable foam sleeping mats obtained from Defense Reutilization and Marketing Office.

Tools

The following tools are needed to make a pugil stick:

Tape measure, ruler, or yardstick.
Scissors.
Grommet machine.
Knife.
Sewing machine.

Directions

To make a pugil stick, Marines perform the following steps.

Preparing Materials

Step 1. Cut foam sleeping mat into the dimensions listed above. Cut mats to double thickness.

Step 2. Use spray adhesive to adhere the two 9" by 27" pieces of foam together. Use the spray adhesive to adhere the two 11" by 19" pieces of foam together. Let adhesive dry before using the double-thick foam for handguards.

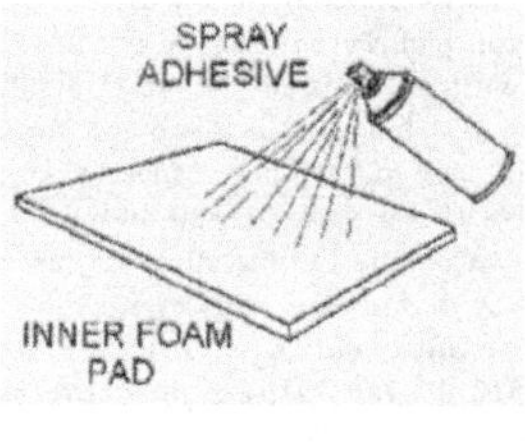

Step 3. Cut canvas material to make two coverings for the inner handguards. Cut material to make two coverings for the outer handguards. Two inner handguards and two outer handguards are required for each pugil stick.

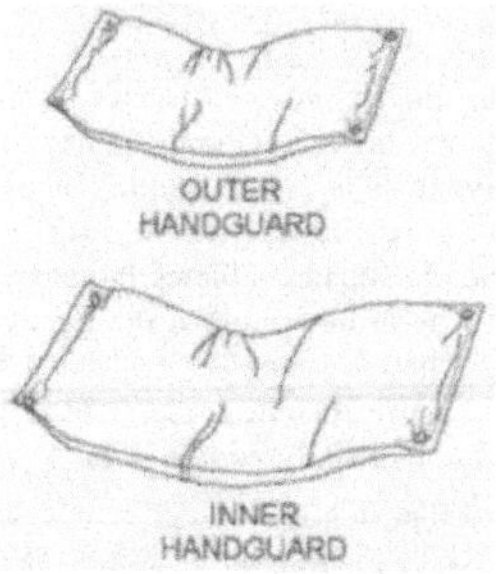

Step 4. Use a sewing machine to make a sleeve out of the 9" by 27" canvas. Sew down three sides. Insert foam pad inside the canvas sleeve and then sew down the fourth side. Use the grommet machine to create holes reinforced with metal grommets in each of the four corners. Once finished, the inner handguards are complete.

Note: It is recommended that the Fabric Repair Shop be used to sew canvas sleeves and apply grommets.

Step 5. Repeat step 4 to make the other inner handguard and both outer handguards.

Constructing Ends, End Plugs, and End Caps

Step 1. Apply spray adhesive on approximately 8 inches of one end of the dowel. This adhesive serves to glue the foam pad to the wooden dowel.

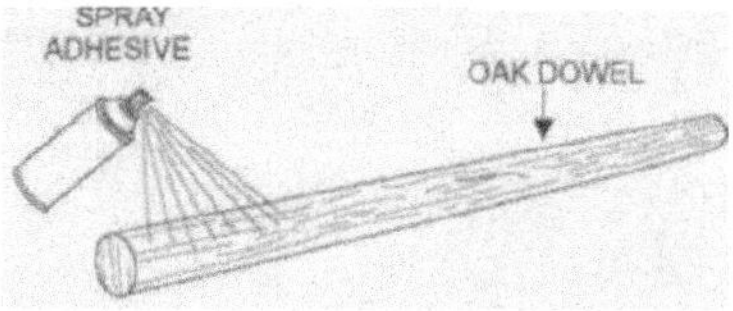

Step 2. Apply spray adhesive on one side of the 11" by 54" foam pad. This causes the foam pad to stick to itself when it is rolled around the end of the wooden dowel. Ensure 3 inches of the rolled foam pad extends past the wooden dowel for stability.

Step 3. Apply spray adhesive on one side of 3" by 14 1/2" foam pad. Roll pad into a tight cylinder.

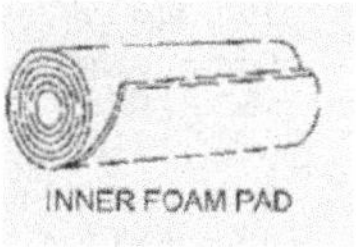

Step 4. Apply spray adhesive on the exposed end of the dowel and fill the hole in the end with the end plug. The end plug prevents the wooden dowel from protruding beyond the edge of the rolled foam padding. This is an important safety precaution to prevent injury.

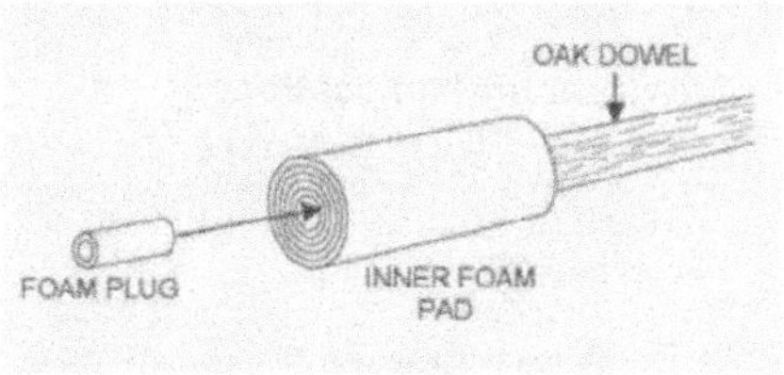

Step 5. Use duct tape or riggers tape to secure and reinforce the foam pad once the end plug is inserted in the end.

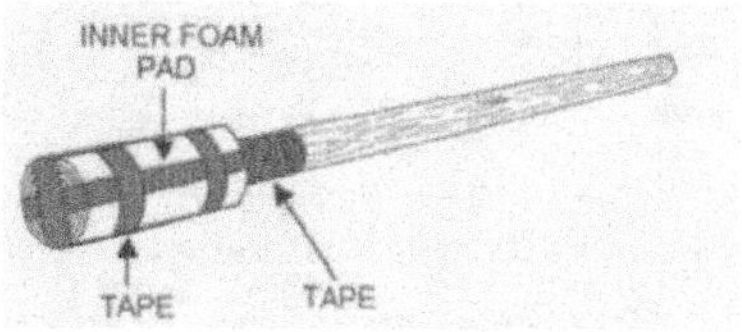

Step 6. Measure the dimensions of the ends (typically, the length of the cylinder should be 12 inches long to allow for taping). Cut and sew canvas material in the shape of a cylinder with one open end. Once complete, this becomes the end cap. The end cap is a colored sleeve that covers the rolled and taped foam pad at each end of the pugil stick.

Note: It is recommended that the Fabric Repair Shop be used to make the end cap.

Step 7. Slide one of the end caps over the rolled foam pad and tape it to the dowel.

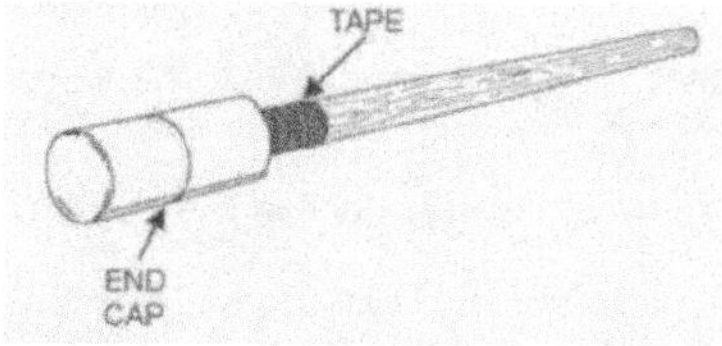

Building the Centerguard

Step 1. Spray one side of the 7" by 13" foam pad with spray adhesive and wrap it tightly around the center of the dowel.

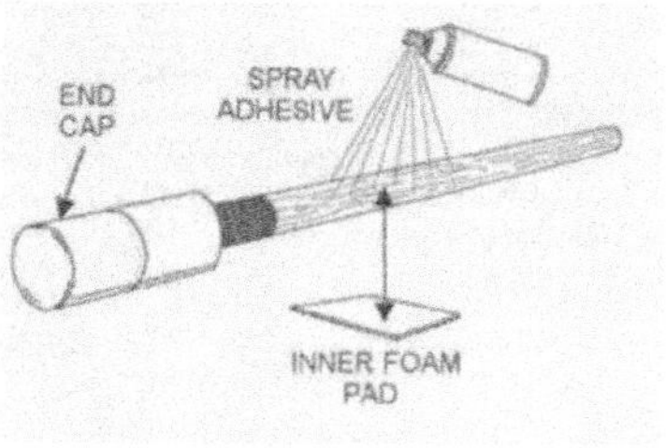

Step 2. Wrap duct tape or riggers tape around the foam pad to secure.

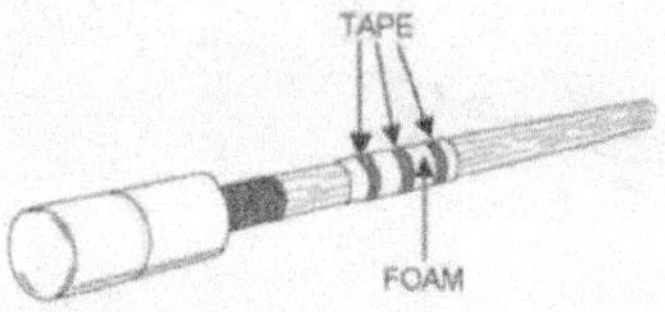

Step 3. Use a sewing machine to make a tubular pugil stick sock out of the 8" by 14" piece of canvas. Slide the pugil stick sock over the rolled foam pad and secure it with the duct tape or riggers tape. Duct tape or riggers tape protects the edges of the centerguard and reinforces the gluing effects of the spray adhesive.

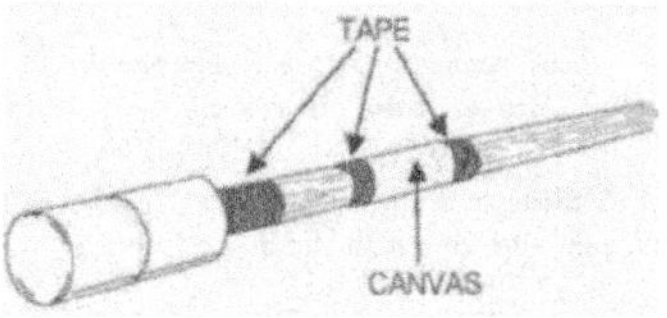

Note: It is recommended that the Fabric Repair Shop be used to make the pugil stick sock.

Finishing the Opposite End Cap

After one end cap is complete and the centerguard is finished, then repeat steps 1 through 7 in Constructing Ends, End Plugs, and End Caps for the other end of the pugil stick. Ensure that the opposite color end cap is used.

Fastening Handguards

Step 1. Place the inner handguards (longer piece) around the dowel. The inner handguard is placed on first because it protects the wrist. One inner handguard is needed for the left hand and one for the right hand.

Step 2. Fold the inner handguard around the pugil stick and run the para-cord or 5-50 cord through the metal grommets.

Step 3. Tighten down the corner ends of the inner handguards by pulling on the para-cord or 5-50 cord.

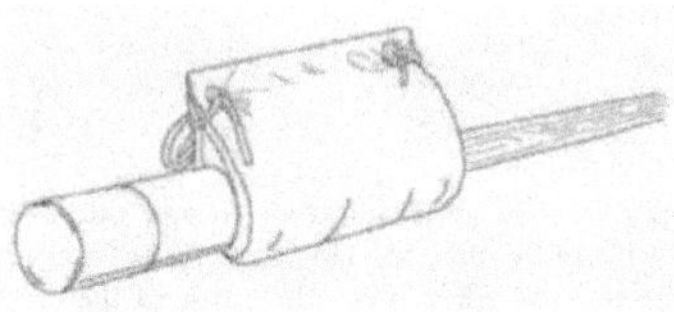

Step 4. Tie off the cord with a square knot above the metal grommets.

Step 5. Place the outer handguards (shorter piece) over the inner handguards (longer piece). Repeat steps 1-4 for the other handguard. The outer handguards provide additional protection to the knuckles and fingers and provides additional stability for the inner handguards.

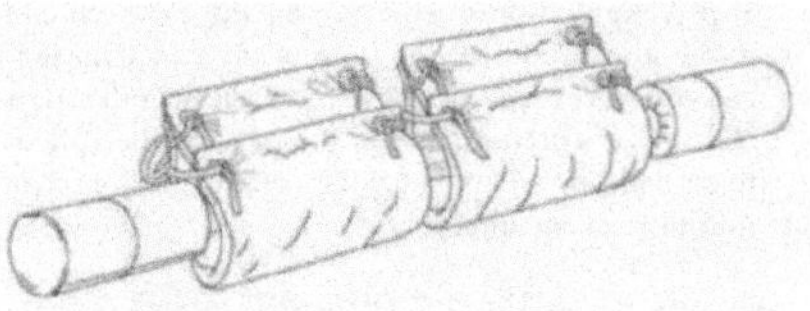

Serviceability Inspection

Prior to any training, all pugil sticks are inspected to ensure they are fit for use. If any of the following conditions exist, the pugil stick is unsafe and will not be used:

- Stick protrudes through the end caps.
- Frayed or worn edges exist on canvas surfaces.
- The stick can be felt through the pads or the padding is too soft.
- End caps or handguards are not securely attached.

(reverse blank)

APPENDIX B

SAFETY PRECAUTIONS DURING TRAINING

1. General Safety Precautions

When training close combat techniques, certain safety precautions must be adhered to to prevent injuries.

Most training should be conducted on a training area with soft footing such as a sandy or grassy area. If training mats are available, they should be used. A hard surface area is not appropriate for close combat training.

All techniques should be executed slowly at first. Marines can increase the speed of execution as they become more proficient. Marine-on-Marine training that requires contact (chokes, throws, ground fighting, and unarmed restraints and manipulation) should not be executed at full force or full speed.

If a technique is applied to the point that a Marine is uncomfortable, the Marine must "tap out." This indicates immediate release of the pressure being applied or to immediately stop the technique. The Marine "taps out" by firmly tapping his hand several times on any part of the opponent's body that will get his attention or by saying *stop*.

Second Impact Syndrome occurs when a second concussion develops within hours, days, or weeks following a prior concussion (and before recovery from the first concussion). Second Impact Syndrome causes rapid brain swelling and can cause death. Marines who experience headaches or other symptoms following training must be examined by appropriate medical personnel. These symptoms can include, but are not limited to, blurred vision, ringing in the ears, dialation of the pupils, bleeding from the ears or mouth, slurred speech, swelling in head or neck area, or any unnatural discoloration of head or neck. They should not be allowed to participate in pugil stick training or any other activity where a heavy blow might be sustained for a minimum of 7 days after the symptoms have subsided.

2. Safety Precautions for Individual Techniques

Falls

When training Marines to fall, they should progress from the ground, to a kneeling or squatting position, and then to a standing position. This ensures they are comfortable and understand the technique before progressing to executing falls from a higher profile. This instructional technique vastly reduces the risk of injury.

Strikes and Punches

When training Marines to strike and punch, they begin by executing the techniques "in the air." As they become more proficient, they execute strikes on equipment (when available) such as an air shield or a heavy bag. At no time should they be permitted to execute strikes on another student.

Chokes and Ground Fighting Chokes

When training Marines to execute chokes, they will not apply pressure to the opponent's throat during training because the trachea and windpipe can be crushed. During training, Marines should practice the proper procedures for blood chokes. They should never execute air chokes. At no time during training should a choke be applied at full force or full speed. No choke should be held for longer than 5 seconds.

Unarmed Restraints and Manipulation

When training Marines to execute unarmed restraints and manipulation techniques, they utilize slow and steady pressure. Never apply these techniques at full force or at full speed.

Pugil Stick Training

A blow to the head during training will bruise the brain. A second blow to the head can cause death. This is known as the Second Impact Syndrome. There must be 7 days between pugil stick training to prevent injury or death.

Table of Contents

Chapter 1. Introduction to Rifle Marksmanship

Chapter 2. Introduction to the M16A2 Service Rifle

Chapter 3. Weapons Handling

Chapter 4. Fundamentals of Marksmanship

Chapter 5. Rifle Firing Positions

Chapter 7. Rifle Presentation

Chapter 8. Effects of Weather

Chapter 9. Zeroing

Appendices

CHAPTER 1. INTRODUCTION TO RIFLE MARKSMANSHIP

All Marines share a common warfighting belief: "Every Marine a rifleman." This simple credo reinforces the belief that all Marines are forged from a common experience, share a common set of values, and are trained as members of an expeditionary force in readiness. As such, there are no "rear area" Marines, and no one is very far from the fighting during expeditionary operations. The Marine rifleman of the next conflict will be as in past conflicts: among the first to confront the enemy and the last to hang his weapon in the rack after the conflict is won.

1001. ROLE OF THE MARINE RIFLEMAN

Marine Corps forces are employed across the entire range of military operations. At one end is war, which is characterized by large-scale, sustained combat operations. At the other end of the scale are those actions referred to as military operations other than war (MOOTW). MOOTW focuses on deterring aggression, resolving conflict, promoting peace, and supporting civil authorities. These operations can occur before, during, and after combat operations. Training and preparation for MOOTW should not detract from the Corps' primary mission of training Marines to fight and win in combat. MOOTW normally does not involve combat. However, Marines always need to be prepared to protect themselves and respond to changing threats and unexpected situations. Whenever the situation warrants the application of deadly force, the Marine rifleman must be able to deliver well aimed shots to eliminate the threat. Sometimes the need for a well aimed shot may even be heightened by the presence of noncombatants in close proximity to the target. The proficient rifleman handles this challenge without unnecessarily escalating the level of violence or causing unnecessary collateral damage. The Marine rifleman must have the versatility, flexibility, and skills to deal with a situation at any level of intensity across the entire range of military operations.

To be combat ready, the Marine must be skilled in the techniques and procedures of rifle marksmanship and take proper care of his rifle. Even when equipped with the best rifle in the world, a unit with poorly trained riflemen cannot be depended upon to accomplish their mission. Usually, poorly trained riflemen either fail to fire their weapon or they waste ammunition by firing ineffectively. To send Marines into harm's way without thorough training in the use of their individual weapons carries undue risks for every Marine in the unit. On the other hand, well trained riflemen can deliver accurate fire against the enemy under the most adverse conditions. A well trained rifleman is not only confident that he can help his unit accomplish it's mission, he is confident that he can protect his fellow Marines and himself.

1002. CONDITIONS AFFECTING MARKSMANSHIP IN COMBAT

Many factors affect the application of marksmanship in combat; among them are—

- Most targets are linear in nature and will consist of a number of men or objects irregularly spaced along covered or concealed areas.
- Most targets can be detected by smoke, flash, dust, noise or movement, but will only be visible for a brief moment before taking cover.
- The nature of the target, irregularities of terrain and vegetation will often require a rifleman to use a position other than prone in order to fire effectively on the target.
- The time in which a target can be engaged is often fleeting.

1003. COMBAT MINDSET

In a combat environment, the Marine must be constantly prepared for possible target engagement. When a target presents itself in combat, there may be very little time to take action. A Marine must be able to engage the target quickly and accurately. The unique demands of combat (i.e., stress, uncertainty) dictate that the Marine be both physically and mentally prepared to engage enemy targets. It will not be enough to simply know marksmanship techniques and procedures. The Marine must develop the mental discipline to prepare for enemy contact. In the confusion, noise, and stress of the combat environment, the Marine must have the ability to eliminate any hesitation, fear or uncertainty of action and to focus on the actions required to fire well-aimed shots. This is accomplished through establishment of a combat mindset. The key factors in the development of a combat mindset include both physical and mental preparation.

Physical Preparation

In combat, targets can present themselves without warning. Therefore, it is essential for the Marine to maintain proper balance and control of his weapon at all times so he can quickly assume a firing position, present the weapon, and accurately engage the target. However, speed alone does not equate to effective target engagement. The Marine should fire only as fast as he can fire accurately, never exceeding his physical ability to apply the fundamentals of marksmanship. To be effective in combat, the Marine must train to perfect the physical skills of shooting so those skills become second nature. Mastery of physical skills allow the Marine to concentrate on the mental aspects of target engagement; e.g., scanning for targets, detection of targets, selection and use of cover. The more physical skills that a Marine can perform automatically, the more concentration he can give to the mental side of target engagement.

Mental Preparation

While combat is unpredictable and constantly changing, the Marine can prepare himself mentally for confrontation with the enemy. The stress of battle, coupled with the often limited time available to engage targets, requires concentration on the mental aspects of target engagement; e.g., scanning for targets, detection of targets, and the selection and use of cover.

Knowledge of the Combat Environment

The Marine must be constantly aware of the surroundings to include the terrain, available cover, possible areas of enemy contact, backdrop of the target, etc. This awareness will enable the Marine to select and assume a firing position and to quickly and accurately engage targets.

Plan of Action

In combat, the situation will dictate the action to be taken. The Marine must understand the situation, identify and evaluate possible courses of action, and then develop a plan for target engagement that accomplishes the mission.

Confidence

The Marine must believe in his ability to engage targets accurately in any combat situation. A Marine's level of confidence is rooted in the belief that future challenges will be overcome—particularly the challenge of firing well aimed shots in a combat environment where the enemy may be returning fire. A key factor in a Marine's level of confidence is the degree to which he has mastered the techniques and procedures of the rifle marksmanship. Mastery of rifle marksmanship can only be obtained by quality instruction. Quality instruction is the foundation for practical application of the marksmanship fundamentals during range and field firing.

CHAPTER 2. INTRODUCTION TO THE M16A2 SERVICE RIFLE

Note

+The procedures in this manual are written for right-handed Marines; left-handed Marines should reverse instructions as necessary.

2001. DESCRIPTION

The M16A2 service rifle is a lightweight, 5.56 millimeter (mm), magazine-fed, gas-operated, air-cooled, shoulder-fired rifle. (Fig. 2-1 shows a right-side view and fig. 2-2, on page 2-2, shows the left-side view.) The rifle fires in either semiautomatic (single-shot) mode or a three-round burst through the use of a selector lever. The M16A2 service rifle has a maximum effective range of 550 meters for individual or point targets. The bore and chamber are chrome-plated to reduce wear and fouling. The handguards are aluminum-lined and are vented to permit air to circulate around the barrel for cooling purposes and to protect the gas tube. An aluminum receiver helps reduce the overall weight of the rifle. The trigger guard is equipped with a spring-loaded retaining pin that, when depressed, allows the trigger guard to be rotated out of the way for access to the trigger while wearing heavy gloves. An ejection port cover prevents dirt and sand from getting into the rifle through the ejection port. The ejection port cover should be closed when the rifle is not being fired. It is automatically opened by the action of the bolt carrier. The muzzle compensator serves as a flash suppressor and assists in reducing muzzle climb.

2002. OPERATIONAL CONTROLS

Selector Lever

The selector lever has three settings—safe, semi, and burst. The setting selected depends on the firing situation. See figure 2-3 on page 2-2.

Safe

The selector lever in the safe position prevents the rifle from firing.

Semi

The selector lever in the semi position allows one shot to be fired with each pull of the trigger.

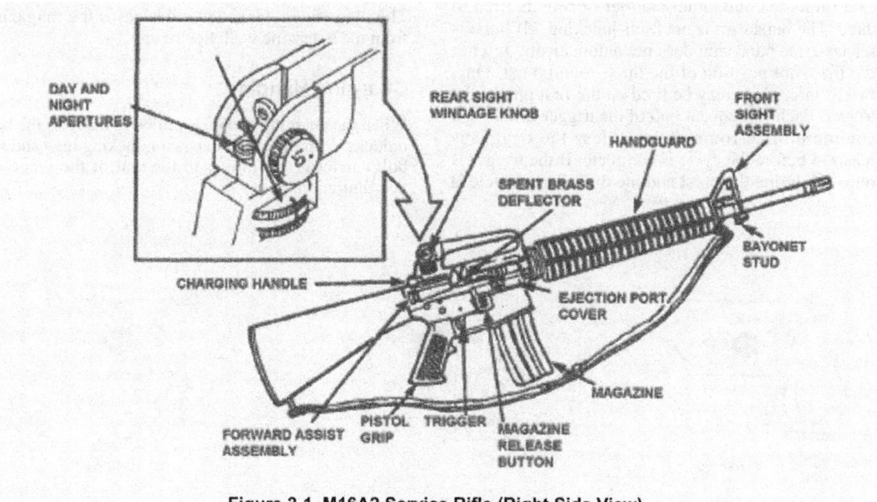

Figure 2-1. M16A2 Service Rifle (Right Side View).

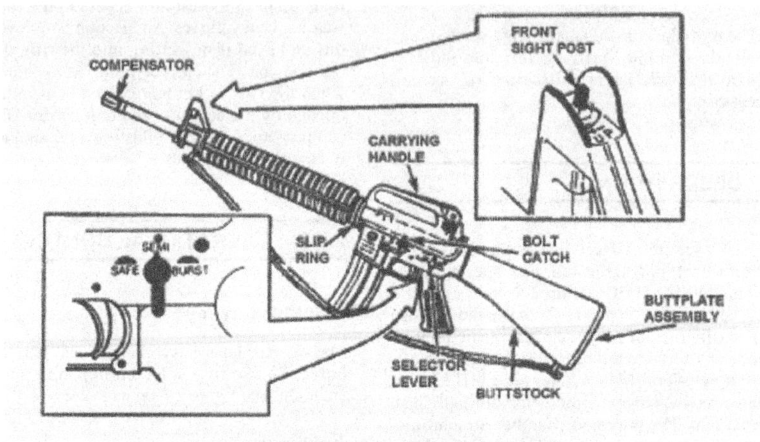

Figure 2-2. M16A2 Service Rifle (Left Side View).

Burst

The selector lever in the burst position allows the rifle to continue its cycle of operation until interrupted by the burst cam. With each pull of the trigger, the burst cam limits the maximum number of rounds fired to three. The burst cam is not "self-indexing." If burst is selected, the burst cam does not automatically reset to the first shot position of the three-round burst. One, two or three shots may be fired on the first pull of the trigger. Each subsequent pull of the trigger results in a complete three-round burst unless the trigger is released before the cycle is complete. If the trigger is released during the burst and the three-round cycle is interrupted, the next pull of the trigger fires the rounds remaining in the interrupted three-round cycle.

Magazine Release Button

The magazine release button releases the magazine from the magazine well. See figure 2-4.

Charging Handle

When the charging handle is pulled to the rear, the bolt unlocks from the barrel extension locking lugs and the bolt carrier group moves to the rear of the receiver. See figure 2-5.

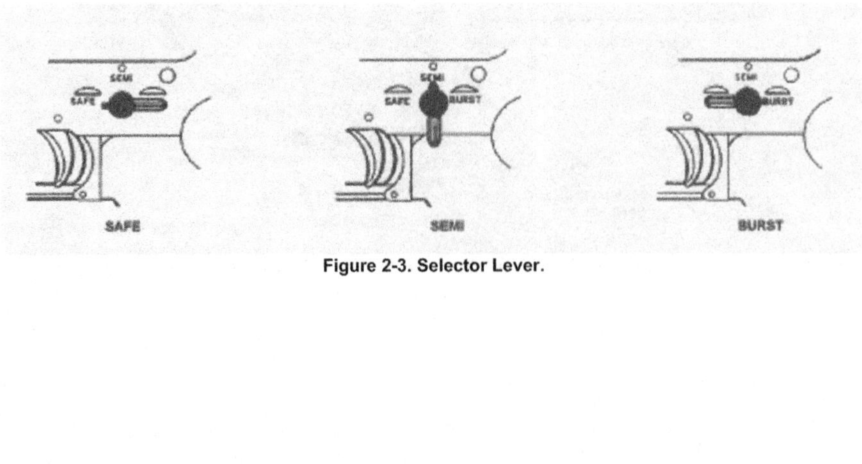

Figure 2-3. Selector Lever.

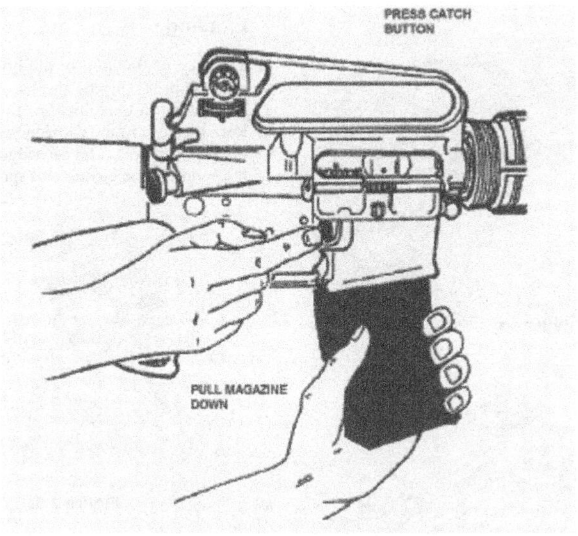

Figure 2-4. Magazine Release Button.

Bolt Catch

If the charging handle is pulled to the rear when the lower portion of the bolt catch is depressed, the bolt carrier group will lock to the rear. When the bolt carrier group is locked to the rear and the upper portion of the bolt catch is depressed, the bolt carrier group will slide forward, driven by the buffer assembly and action spring, into the firing position.

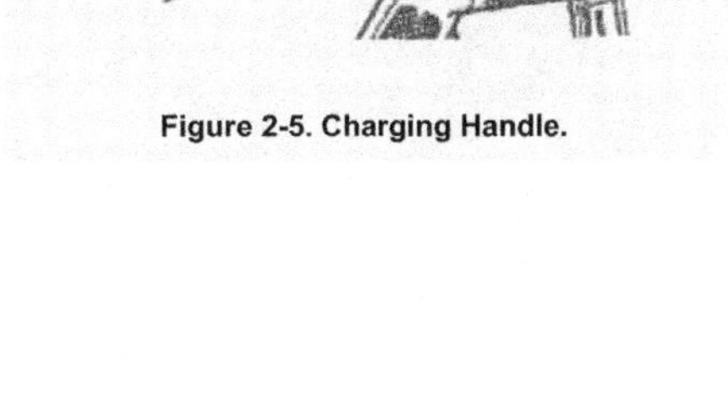

Figure 2-5. Charging Handle.

2003. CYCLE OF OPERATION

Firing

The hammer releases and strikes the head of the firing pin, driving the firing pin into the round's primer. The primer ignites the powder in the cartridge. Gas generated by the rapid burning of powder propels the projectile through the barrel. After the projectile passes the gas port, a portion of the expanding gas enters the gas port and gas tube. The gas tube directs the gas rearward into the bolt carrier key and causes the bolt carrier to move rearward. See figure 2-6 on page 2-4.

Unlocking

Figure 2-7 on page 2-4 illustrates unlocking of the bolt. As the bolt carrier moves to the rear, the bolt cam pin follows the path of the cam track located in the bolt carrier. This causes the bolt assembly to rotate until the bolt-locking lugs are no longer aligned behind the barrel extension locking lugs.

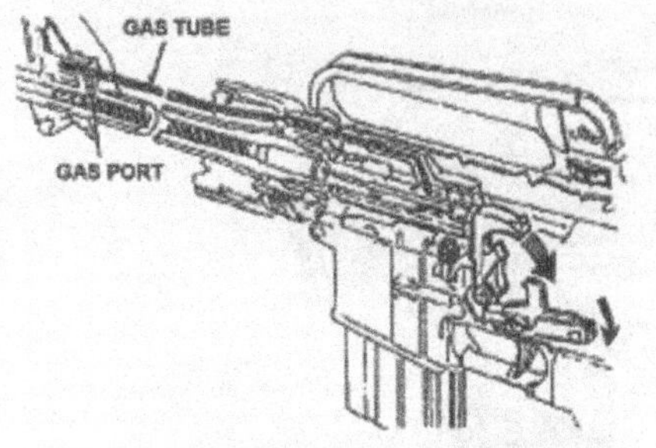

Figure 2-6. Firing.

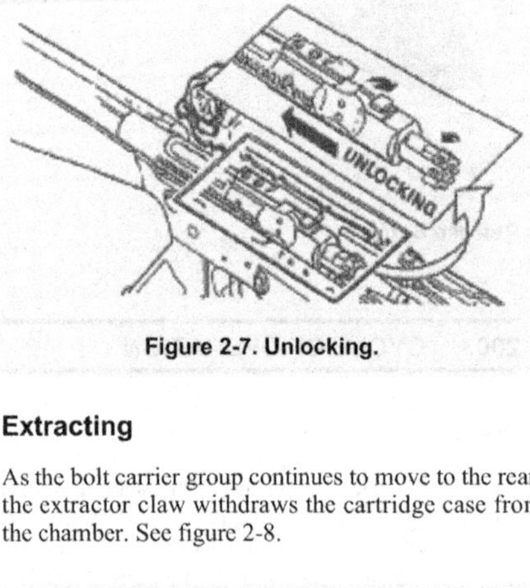

Figure 2-7. Unlocking.

Extracting

As the bolt carrier group continues to move to the rear, the extractor claw withdraws the cartridge case from the chamber. See figure 2-8.

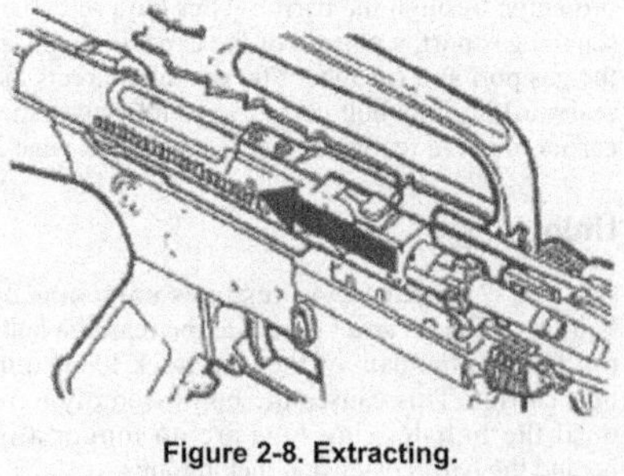
Figure 2-8. Extracting.

Ejecting

The ejector, located in the bolt face, is compressed into the bolt body by the base of the cartridge case. The rearward movement of the bolt carrier group allows the nose of the cartridge case to clear the front of the ejection port. The cartridge case is thrown out by the action of the ejector and spring. See figure 2-9.

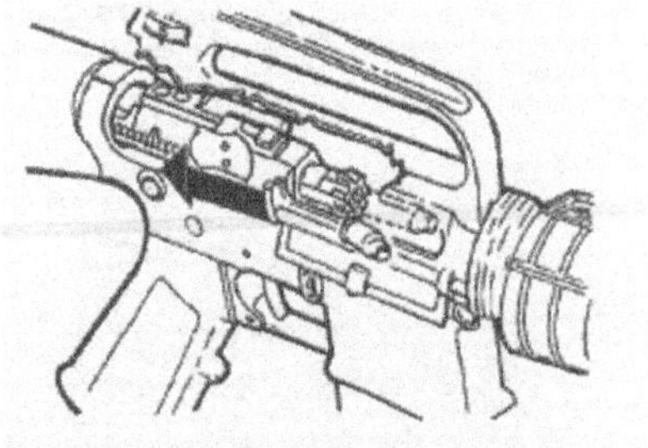
Figure 2-9. Ejecting.

Cocking

Continuing its rearward travel, the bolt carrier overrides the hammer, forces it down into the receiver, compresses the hammer spring, and causes the disconnector to engage the lower hammer hook. See figure 2-10.

Figure 2-10. Cocking.

Feeding

Once rearward motion causes the bolt carrier group to clear the top of the magazine, the expansion of the magazine spring forces a round into the path of the bolt. After the action spring overcomes and absorbs the rearward motion of the bolt carrier group, it expands and sends the buffer assembly and bolt carrier group forward with enough force to strip a round from the magazine. See figure 2-11.

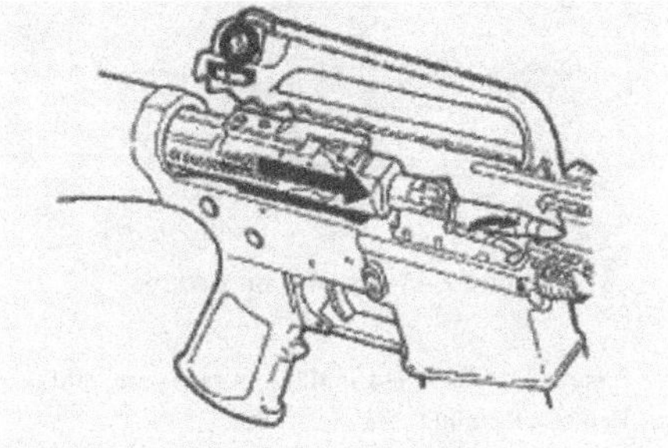

Figure 2-11. Feeding.

Chambering

As the bolt carrier group continues to move forward, pushing a fresh round in front of it, the face of the bolt thrusts the new round into the chamber. The extractor claw grips the rim of the cartridge case. The ejector is forced into its hole, compressing the ejector spring. See figure 2-12.

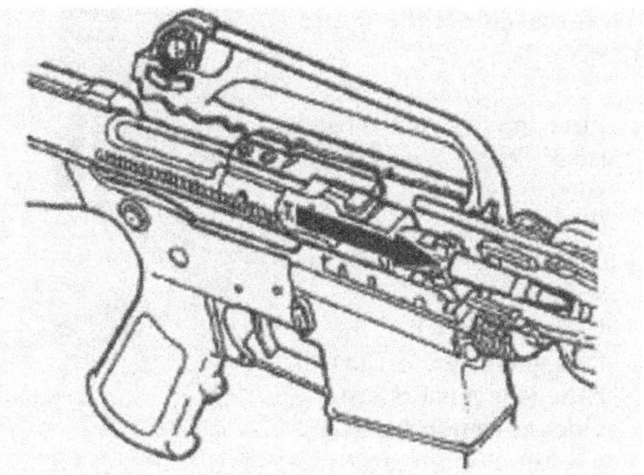

Figure 2-12. Chambering.

Locking

As the bolt carrier group continues to move forward, the bolt-locking lugs are forced against the barrel extension and the bolt cam pin is forced along the cam track. The bolt rotates and aligns the bolt locking lugs behind the barrel extension locking lugs. The weapon is ready to fire. See figure 2-13.

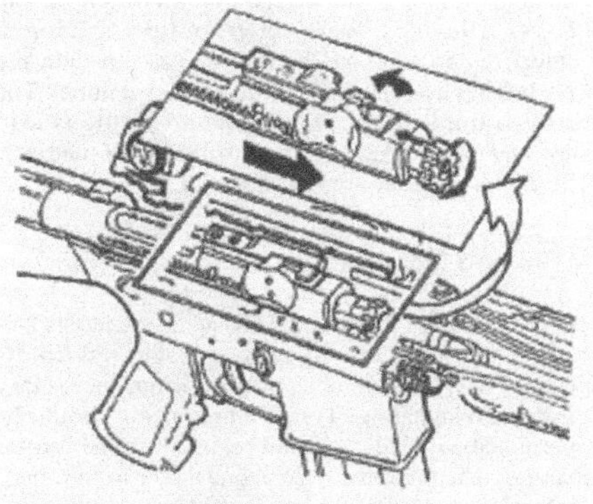

Figure 2-13. Locking.

2004. Ammunition

Four types of ammunition are authorized for use with the M16A2 service rifle: ball (M193 and M855), tracer (M196 and M856), dummy (M199), and blank (M200) (see fig. 2-14).

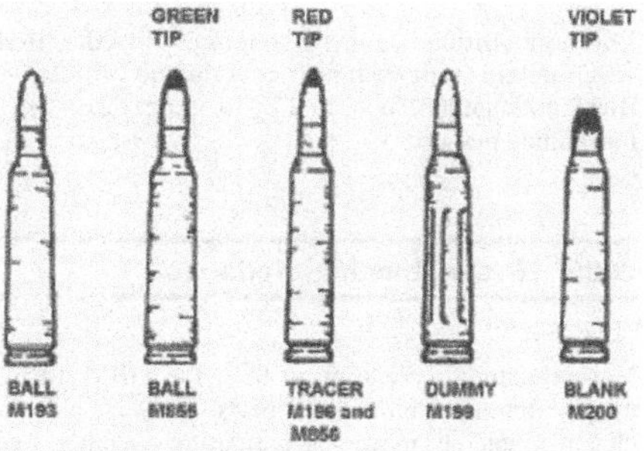

Figure 2-14. Authorized Ammunition.

M193 Ball

This ammunition is a 5.56mm centerfire cartridge with a 55-grain gilded-metal jacket, lead alloy core bullet. The primer and case are waterproofed. The M193 ball ammunition has no identifying marks.

M855 Ball

This ammunition is the primary ammunition for the M16A2 rifle. Identified by a green tip, its 5.56mm centerfire cartridge has better penetration than the M193. It has a 62-grain gilded-metal jacket bullet. The rear two-thirds of the core of the projectile is lead alloy and the front one-third is a solid steel penetrator. The primer and case are waterproofed.

M196 and M856 Tracer

This ammunition has the same basic characteristics as ball ammunition. Identified by a bright red tip, its primary uses include observation firing, incendiary effect, and signaling. Tracer ammunition should be intermixed with ball ammunition in a ratio no greater than 1:1. The preferred ratio is one tracer to four balls (1:4) to prevent metal fouling in the bore.

M199 Dummy

This ammunition has six grooves along the side of the case. It contains no propellants or primer. The primer well is open to prevent damage to the firing pin. The dummy cartridge is used during dry fire and other training purposes.

M200 Blank

This ammunition has the case mouth closed with a seven-petal rosette crimp. It contains no projectile. Blank ammunition, identified by its violet tip, is used for training purposes.

2005. Preventive Maintenance

Normal care and cleaning of the rifle will result in proper functioning of its all parts. Only issue-type cleaning materials maybe used. Improper maintenance can cause stoppages, reducing combat readiness and effectiveness.

Main Group Disassembly

The M16A2 service rifle is disassembled into three main groups (see fig. 2-15). Before disassembling the M16A2 service rifle—

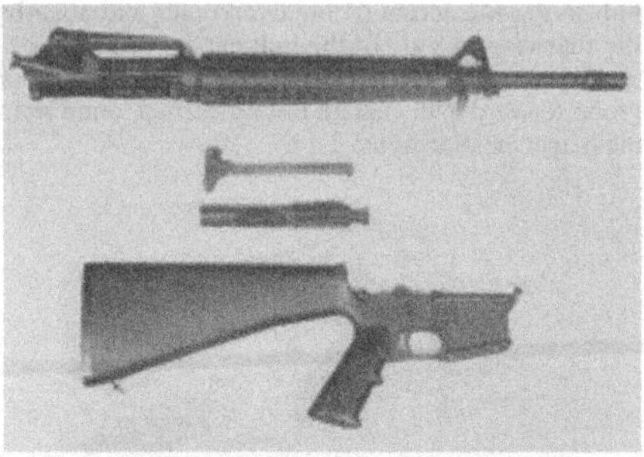

Figure 2-15. Three Main Groups.

- Ensure the rifle is in Condition 4 (see para. 3002).
- Remove the sling.

Note

To facilitate control and ease of disassembly, the handguards may be removed using the "buddy system" as described in Technical Manual (TM) 05538C-10/1A, *Operator's Manual w/Components List Rifle, 5.56-mm, M16A2E W/E (1005-01-128-9936).*

Upper Receiver

To disassemble the upper receiver—

- Place the rifle on the buttstock. Press down on the slip ring with both hands. Pull the handguards free (see fig. 2-16).

  Use caution when the handguards are off the rifle. Handguards provide protection for the gas tube. Damage to the gas tube adversely affects the functioning of the rifle.

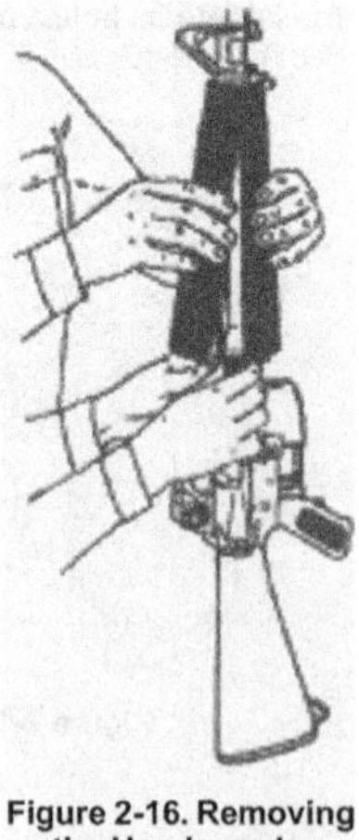

Figure 2-16. Removing the Handguards.

- Move the take down pin from the left to the right as far as it will go to allow the lower receiver to pivot down from the upper receiver.
- Move the receiver pivot pin from left to right as far as it will go and separate the upper and lower receivers.
- Pull back the charging handle and bolt carrier about 3 inches and remove the bolt carrier group.
- Remove the charging handle by sliding it back and down, out of the upper receiver.

No further disassembly is conducted on the upper receiver group.

Bolt Carrier

To disassemble the bolt carrier—

- Remove the firing pin retaining pin.
- Push the bolt back into the bolt carrier to the locked position.
- Tap the base of the bolt carrier against the palm of your hand so the firing pin will drop out.
- Rotate the bolt cam pin one-quarter turn and lift the bolt cam pin out.
- Withdraw the bolt assembly from the carrier.
- Press on the extractor's rear and use the firing pin to push out the extractor-retaining pin. Remove the extractor and spring (the spring is permanently attached to the extractor). (See fig. 2-17.)

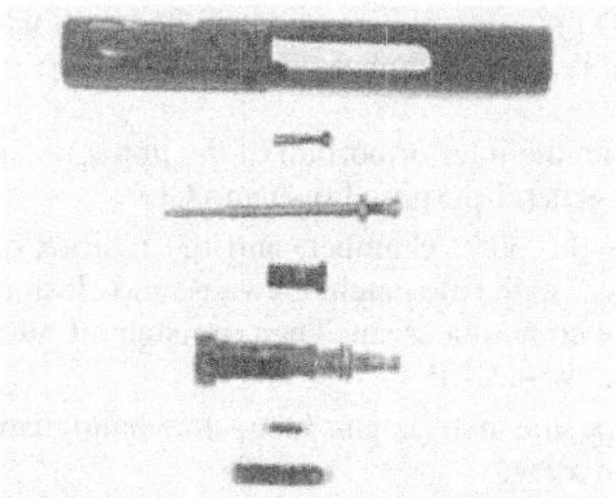

Figure 2-17. Bolt Carrier Disassembled.

CAUTION

Be careful not to damage the tip of the firing pin while pushing out the extractor-retaining pin.

Note

The extractor assembly has a rubber insert within the spring. Do not attempt to remove it. If the spring comes loose, put the large end of the spring in the extractor and seat it. Push in the extractor pin.

Lower Receiver

To disassemble the lower receiver—

- Press in the buffer and depress the buffer retainer.

Note

It may be necessary to use the edge of the charging handle to depress the buffer retainer.

- Press the hammer downward and ease the buffer and action spring forward and out of the receiver.
- Separate the parts. See figure 2-18.

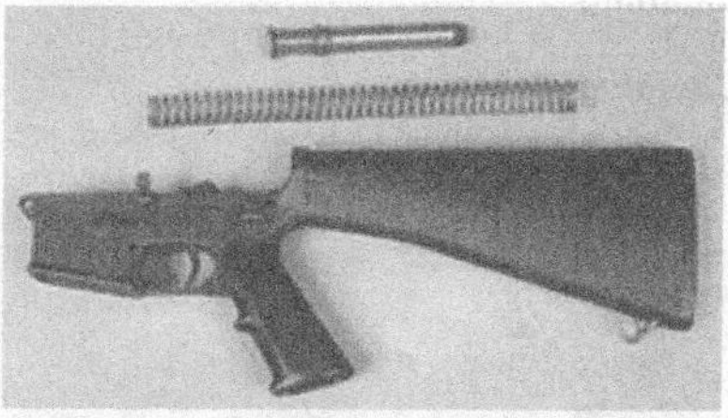

Figure 2-18. Lower Receiver Disassembled.

No further disassembly of the lower receiver is performed.

Note

In combat situations, the rifle may be partially disassembled in any sequence. However, combat situations are the exception, not the rule. Under normal operating circumstances, disassemble the rifle in the sequence just performed. Any further disassembly of the rifle is to be performed by a qualified armorer.

Magazine Disassembly

The magazine should be disassembled regularly for cleaning to avoid the possibility of malfunction or stoppage of the rifle caused by dirty or damaged magazines. To disassemble the magazine—

- Pry up and push base plate out from the magazine.
- Jiggle the spring and follower to remove. Do not remove the follower from the spring. See figure 2-19.

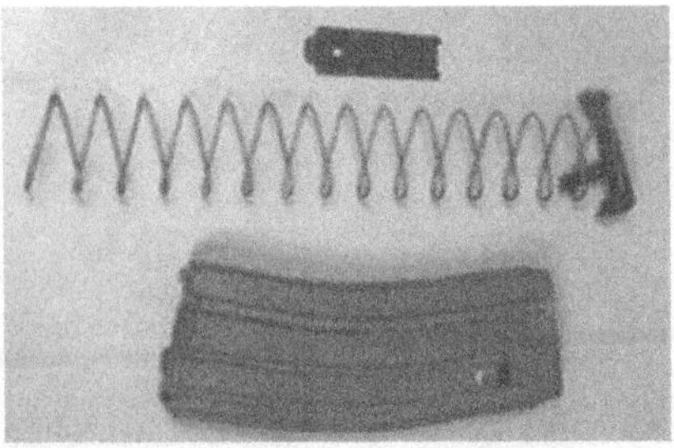

Figure 2-19. Magazine Disassembled.

Cleaning

Cleaning Materials

The following cleaning materials are used in preventive maintenance (see fig. 2-20):

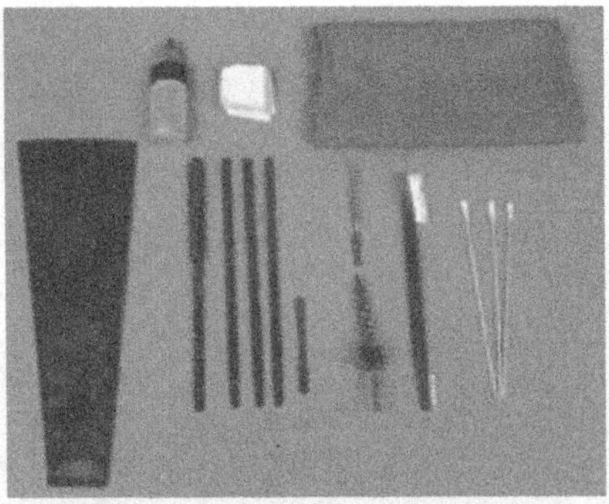

Figure 2-20. Cleaning Materials.

- Cleaner, lubricant, and preservative (CLP). Always shake the bottle well before use.
- Rod in three sections and a handle assembly.
- Patch holder section, swabs, patches, pipe cleaners, and clean rags.
- Brushes: bore, chamber, and general purpose.

Cleaning the Upper Receiver

Basic cleaning of the upper receiver group should include the following:

- Attach the three rod sections together but leave each one about two turns short of being tight.
- Attach the patch holder onto the rod.
- Point the muzzle down and insert the non-patch end of the rod into the chamber. Attach the handle to the cleaning rod section and pull a CLP-moistened 5.56mm patch through the bore.
- Attach the bore brush to the rod but leave it two turns short of being tight. Put a few drops of CLP on the bore brush. Insert the rod into the barrel from the chamber end, attach the handle, and pull the brush through the bore. Repeat 3 times. Remove bore brush and attach the patch holder to the rod with a CLP moistened patch insert the rod into the barrel from the chamber end, attach the handle, and pull the patch through the bore.
- Inspect the bore for cleanliness by holding the muzzle to your eye and looking into the bore.
- Repeat the above steps until the patches come out of the bore clean.
- Attach the chamber brush and one section of the cleaning rod to the handle. Moisten it well with CLP and insert it into the chamber.
- Scrub the chamber and bolt lugs using a combination of a plunging and clockwise rotating action.

Note

Do not reverse direction of the brush while it is in the chamber.

- Clean the interior portion of the upper receiver with the general-purpose brush and CLP.
- Dry the bore, chamber, and the interior of the receiver with rifle patches, swabs, and clean rags until they come out clean. Then moisten all interior surfaces with CLP.
- Wipe the barrel, gas tube, and handguards clean with a rag.

Cleaning the Bolt Carrier Group

- Clean the outer and inner surfaces of the bolt carrier with a general-purpose brush.
- Clean the bolt carrier key with a pipe cleaner.
- Clean the locking lugs, gas rings, and exterior of the bolt with the general-purpose brush.
- Insert a swab into the rear of the bolt and swab out the firing pin recess and gas ports.

- Clean the extractor with the general-purpose brush, ensuring all the carbon is removed from underneath the extractor lip.
- Clean extractor pin, firing pin, and firing pin retaining pin using the general-purpose brush and CLP.
- Clean charging handle assembly with the general-purpose brush and patches.

Cleaning the Lower Receiver

- Wipe dirt from the firing mechanism using a general-purpose brush, clean patch, pipe cleaners, and swabs.
- Clean the outside of the receiver with the general-purpose brush and CLP. Clean the buttplate and rear sling swivel, ensuring drain hole is clear of dirt.
- Wipe the inside of the buffer tube, buffer, and action spring.
- Wipe the inside of the magazine well with a rag.
- Wipe out the inside of the pistol grip and ensure that it is clean.

Cleaning the Magazine

- Clean the inside of the magazine with the general-purpose brush and CLP.
- Wipe dry.
- Keep the spring lightly oiled.

Inspection

While cleaning the rifle, and during each succeeding step in the preventive maintenance process, inspect each part for cracks and chips and ensure parts are not bent or badly worn. Report any damaged part to the armorer. Inspection is a critical step to ensure the combat readiness of your rifle. It is performed normally during rifle cleaning (prior to lubrication), however, it can be performed throughout the preventive maintenance process.

Lubrication

Lubrication is performed as part of the detailed procedure for preventive maintenance. Lubrication procedures are also performed in preparation for firing.

Lubricant

In all but the coldest arctic conditions, CLP is the lubricant for the rifle. Remember to remove excess CLP from the bore and chamber before firing.

- Lightly lube means that a film of CLP barely visible to the eye should be applied.
- Generously lube means the CLP should be applied heavily enough that it can be spread with the finger.

Upper Receiver

- Lightly lube the inside of the upper receiver, bore, chamber, outer surfaces of the barrel, and surfaces under the handguard.
- Depress the front sight detent and apply two or three drops of CLP to the front sight detent. Depress several times to work lubrication into the spring.
- Lubricate the moving parts and elevation screw shaft of the rear sight.

Bolt Carrier Group

- Generously lube the outside of the cam pin area, the bolt rings, and outside the bolt body.
- Lightly lube the charging handle and the inner and outer surfaces of the bolt carrier.

Lower Receiver

- Lightly lube the inside of the lower receiver extension.
- Generously lube the moving parts inside the lower receiver and their pins.

Reassembly

Reassembling the Rifle

- Return all cleaning gear into the buttstock of the rifle and close the buttplate.
- Connect the buffer and action spring and insert them into the buffer tube/stock.
- Place the extractor and spring back on the bolt. Depress the extractor to align the holes and reinsert the extractor pin.
- Insert the bolt into the carrier. Do not switch bolts between rifles.
- Hold the bolt carrier with the bolt carrier key at 12 o'clock. Insert the bolt into the bolt carrier with the extractor at 12 o'clock.

- Rotate the bolt counterclockwise until the cam pinhole aligns to the cam pin slot in the bolt carrier.

WARNING WARNING WARNING WARNING

WARNING Ensure the cam pin is installed in the bolt group or the rifle may explode while firing.

- Insert the bolt cam pin through the bolt carrier and into the bolt. Rotate the cam pin 1/4 turn right or left. Pull the bolt forward until it stops.
- Drop in the firing pin from the rear of the bolt carrier and seat it.
- Replace the firing pin retaining pin. Ensure the head of the firing pin retaining pin is recessed inside the bolt carrier. The firing pin should not fall out when the bolt carrier group is turned upside down.
- Place the charging handle in the upper receiver by lining it up with the grooves in the receiver. Push the charging handle partially in.
- With the bolt in the unlocked position, place bolt carrier key into the groove of the charging handle.
- Push the charging handle and bolt carrier group into the upper receiver until the charging handle locks.
- Join the upper and lower receivers and engage the receiver pivot pin.
- Ensure the selector lever is on safe before closing the upper receiver. Close the upper and lower receiver groups. Push in the takedown pin.
- Install the handguards.
- Attach the sling.

Reassembling the Magazine

To reassemble the magazine—

- Insert the follower and jiggle the spring to install.
- Slide the base under all four tabs until the base catches. Make sure the printing is on the outside.

2006. Function Check

A function check is performed to ensure the rifle operates properly. To perform a function check:

- Place the weapon in Condition 4 (see para. 3002).
- Pull the charging handle to the rear and release. Ensure the selector lever is on safe and pull the trigger. The hammer should not fall.
- Place the selector lever on semi. Pull the trigger and hold it to the rear. The hammer should fall. Pull the charging handle to the rear and release. Release the trigger and pull again. The hammer should fall.
- Pull the charging handle to the rear and release. Place the selector lever on burst. Pull the trigger and hold it to the rear. The hammer should fall. Pull the charging handle to the rear three times and release. Release the trigger and pull again. The hammer should fall.
- Pull the charging handle to the rear and release. Place the selector lever on safe.

2007. User Serviceability Inspection

Individual Marines must perform user serviceability inspections on their weapons before firing them. This inspection ensures the weapon is in an acceptable operating condition.

- Place rifle in Condition 4 (see para. 3002).
- Conduct a function check.
- Check the rifle to ensure the following:
- Compensator is tight.
- Barrel is tight.
- Front sight post is straight.
- Front sight post is adjustable.
- Handguards are serviceable.
- Rear sight elevation and windage knobs are adjustable and have distinct clicks.
- Stock is tight on the lower receiver.
- Weapon is properly lubricated for operational conditions.
- Barrel is clear of obstructions.
- Gas rings are serviceable.
- Ensure magazines are serviceable.
- Load the rifle with an empty magazine. Ensure that the magazine can be seated.
- Without depressing the bolt catch, pull the charging handle to the rear. Ensure that the bolt locks to the rear.
- Depress the upper portion of the bolt catch and observe bolt moving forward on an empty chamber.

Ensure the bolt moves completely forward and locks in the chamber.

- Repeat this procedure with all magazines.

2008. Field Maintenance

Preventive maintenance in the field is performed when detailed disassembly and cleaning is not practical due to operational tempo or the level of threat. To perform limited field preventive maintenance—

- Place the rifle in Condition 4 (see paragraph 3002).
- Break the rifle down by removing the rear take down pin and rotating the upper receiver and barrel forward.
- Remove the bolt carrier group.
- Do not disassemble the bolt carrier group further.
- Clean the bolt carrier group.
- Clean the upper and lower receiver groups (without further disassembly).
- Clean the bore and chamber.
- Lubricate the rifle.
- Reassemble the rifle and perform a user serviceability inspection.

2009. Cleaning the Rifle in Various Conditions

The climatic conditions in various locations require special knowledge about cleaning and maintaining the rifle. The conditions that will affect the rifle the most are: hot, wet tropical; hot, dry desert; arctic or low temperature; and heavy rain and fording.

Hot, Wet Tropical

- Perform normal maintenance.
- Clean and lubricate your rifle more often. Inspect hidden surfaces for corrosion. Pay particular attention to spring-loaded detentes.
- Use lubricant more liberally.
- Unload and check the inside of the magazine more frequently. Wipe dry and check for corrosion.
- When practical, keep the rifle covered.

Hot, Dry Desert

Hot dry climates are usually areas that contain blowing sand and fine dust. Dust and sand will get into the rifle and magazines, causing stoppages. It is imperative to pay particular attention to the cleaning and lubrication of the rifle in this type of climate.

Corrosion is less likely to form in these environments, and lubrication will attract more dirt. For this reason, use lubrication more sparingly.

Whenever practical, keep the rifle covered.

Arctic or Low Temperature

- Clean and lubricate the rifle in a warm room, with the rifle at room temperature, if possible. Lubricating oil, arctic weapons (LAW) can be used below a temperature of 0 degrees Fahrenheit and must be used below -35 degrees Fahrenheit.
- Keep the rifle covered when moving from a warm to a cold environment to allow gradual cooling of the rifle. This prevents the condensation of moisture and freezing. Condensation will form on the rifle when it is moved from outdoors to indoors. If possible, leave the rifle in a protected but cold area outdoors. When bringing the rifle inside to a warm place, it should be disassembled and wiped down several times as it warms.
- Always try to keep the rifle dry.
- Unload and perform a function check every 30 minutes, if possible, to help prevent freezing of functional parts.
- Do not lay a warm rifle in snow or ice.
- Keep the inside of the magazine and ammunition wiped dry. Moisture will freeze and cause stoppages.

Heavy Rain and Fording

- Keep the rifle dry and covered when practical.
- Keep water out of the barrel if possible. If water does get in, drain and (if possible) dry with a patch. If water is in the barrel, point the muzzle down and break the seal by doing a chamber check so the water will drain. If water is in the stock of the weapon, ensure the drain hole in the stock is clear so the water can run out.
- Perform normal maintenance.

CHAPTER 3. WEAPONS HANDLING

Weapons handling procedures provide a consistent and standardized way for a Marine to handle, operate, and employ the rifle safely and effectively. Proper weapons handling procedures ensure the safety of Marines by eliminating negligent discharges and reinforcing positive identification of targets before engagement. Weapons handling procedures apply at all levels of training and during combat operations.

Note

+The procedures in this manual are written for right-handed Marines; left-handed Marines should reverse instructions as necessary.

3001. Safety Rules

Safe handling of the rifle is critical. If proper weapons handling procedures are not used, a Marine risks his safety and the safety of his fellow Marines. During combat, a Marine must react quickly, safely, and be mentally prepared to engage targets. To ensure that only the intended target is engaged, a Marine must apply the following safety rules at all times:

Rule 1—Treat every weapon as if it were loaded. When a Marine takes charge of a rifle in any situation, he must treat the weapon as if it were loaded, determine its condition (see para. 3003), and continue applying the other safety rules.

Rule 2—Never point a weapon at anything you do not intend to shoot. A Marine must maintain muzzle awareness at all times.

Rule 3—Keep your finger straight and off the trigger until you are ready to fire. A target must be identified before taking the weapon off safe and moving the finger to the trigger.

Rule 4—Keep the weapon on safe until you intend to fire. A target must be identified before taking the weapon off safe. This rule is intended to eliminate the chance of the weapon discharging by accident (e.g., brush snagging the trigger).

3002. Weapons Conditions

A weapon's readiness is described by one of four conditions. The steps in the loading and unloading process take the rifle through four specific conditions of readiness for live fire.

Condition 1. Safety on, magazine inserted, round in chamber, bolt forward, ejection port cover closed.

Condition 2. Not applicable to the M16A2 rifle.

Condition 3. Safety on, magazine inserted, chamber empty, bolt forward, ejection port cover closed.

Condition 4. Safety on, magazine removed, chamber empty, bolt forward, ejection port cover closed.

3003. Determining a Weapon's Condition (Chamber Check)

A Marine must know the condition of his weapon at all times. When a Marine takes charge of a weapon in any situation, he must determine its condition. Situations include coming across an unmanned rifle in combat, taking charge of any weapon after it has been unmanned (e.g., out of a rifle rack, stored in a vehicle), or taking charge of another Marine's weapon. To determine the condition of the weapon in any of these situations, the Marine should:

- Determine if a magazine is present.
- Ensure the rifle is on safe.

- Conduct a chamber check.
- Bring the left hand back against the magazine well.
- Extend the fingers of the left hand and cover the ejection port (see fig. 3-1).

Figure 3-1. Position of Hand.

- Grasp the charging handle with the index and middle fingers of the right hand.
- Pull the charging handle slightly to the rear and visually and physically inspect the chamber (see fig. 3-2). *Right-handed Marines*, insert one finger of the left hand into the ejection port and feel whether a round is present. *Left-handed Marines*, insert the thumb of the right hand into the ejection port and feel whether a round is present.

Figure 3-2. Chamber Check.

Note

The same procedure is used in daylight as during low visibility. A chamber check may be conducted at any time.

CAUTION

Pulling the charging handle too far to the rear while inspecting the chamber may cause double feed or ejection of one round of ammunition.

- Release the charging handle and observe the bolt going forward.
- Tap the forward assist.
- Close the ejection port cover (if time and the situation permit).
- Remove the magazine (if present) and observe if ammunition is present. If time permits, count the rounds. Reinsert the magazine into magazine well.

3004. Weapons Commands

Weapons commands dictate the specific steps required to load and unload the rifle. Six commands are used in weapons handling:

Load. This command is used to take the weapon from Condition 4 to Condition 3.

Make Ready. This command is used to take the weapon from Condition 3 to Condition 1.

Fire. This command is used to specify when a Marine may engage targets.

Cease-Fire. This command is used to specify when a Marine must stop target engagement.

Unload. This command is used to take the weapon from any condition to Condition 4.

Unload and Show Clear. This command is used when an observer must check the weapon to verify that no ammunition is present before the rifle is placed in Condition 4.

Loading the Rifle

Perform the following steps to load the rifle (take the rifle to Condition 3):

- Ensure the rifle is on safe.
- Withdraw the magazine from the magazine pouch.
- Observe the magazine to ensure it is filled.
- Fully insert the magazine in the magazine well. Without releasing the magazine, tug downward on the magazine to ensure it is seated.
- Close the ejection port cover.
- Fasten the magazine pouch.

Making the Rifle Ready

Perform the following steps to make the rifle ready for firing (take the rifle to Condition 1):

- Pull the charging handle to the rear and release. There are two methods of doing this:
- Grip the pistol grip firmly with the right hand and pull the charging handle with the left hand to its rearmost position and release (see fig. 3-3).

  Or grip the handguards firmly with the left hand and pull the charging handle with the right hand to its rearmost position and release (see fig. 3-4).

Figure 3-3. Pulling the Charging Handle with the Left Hand.

Figure 3-4. Pulling the Charging Handle with the Right Hand.

- To ensure ammunition has been chambered, conduct a chamber check (see para. 3003) to ensure a round has been chambered.
- Check the sights (to ensure proper battlesight zero [BZO] setting, correct rear sight aperture, etc.).
- Close ejection port cover.

Fire

On the command "Fire," aim the rifle, take the rifle off safe, and pull the trigger.

Cease-Fire

On the command "Cease Fire," perform the following steps:

- Place your trigger finger straight along the receiver.
- Place the weapon on safe.

Unloading the Rifle

Perform the following steps to unload the rifle (take the rifle to Condition 4):

- Ensure the weapon is on safe.
- Remove the magazine from the rifle and retain it on your person.

- Cup the left hand under the ejection port, rotate the weapon until the ejection port faces down.
- Pull the charging handle to the rear and catch the round in the left hand (see fig. 3-5).

Figure 3-5. Catching the Round.

- Lock the bolt to the rear.
- Put the weapon on safe if the selector lever would not move to safe earlier.
- Ensure the chamber is empty and that no ammunition is present.
- Depress the bolt catch and observe the bolt moving forward on an empty chamber (see fig. 3-6).

Figure 3-6. Observing the Chamber.

- Close the ejection port cover.
- Check the sights (for proper BZO setting, correct rear sight aperture, etc.).
- Place any ejected round into the magazine and return the magazine to the magazine pouch and close the magazine pouch.

Unloading and Showing the Rifle Clear

Perform the following steps to unload the rifle and show it clear to an observer (take the rifle to Condition 4). The Marine—

- Ensures the weapon is on safe.
- Removes the magazine from the rifle and retains it.
- Cups the left hand under the ejection port, rotates the weapon until the ejection port faces down.
- Pulls the charging handle to the rear and catches the round in the left hand.
- Locks the bolt to the rear and ensures the chamber is empty and that no ammunition is present.
- Has another Marine inspect the weapon to ensure no ammunition is present (see fig. 3-7).

Figure 3-7. Observer Inspection.

The observer—

- Visually inspects the chamber to ensure it is empty, no ammunition is present, and the magazine is removed.
- Ensures the weapon is on safe.
- Acknowledges the rifle is clear.

The Marine, after receiving acknowledgment that the rifle is clear—

- Depresses the bolt catch and observes the bolt moving forward on an empty chamber.
- Closes the ejection port cover.
- Checks the sights (for proper BZO setting, correct rear sight aperture, etc.).
- Places any ejected round into the magazine and returns the magazine to the magazine pouch and closes the magazine pouch.

3005. Filling, Stowing, and Withdrawing Magazines

Filling the Magazine with Loose Rounds

Perform the following steps to fill the magazine:

- Remove a magazine from the magazine pouch.
- Place a round on top of the follower.
- Press down until the round is held between the follower and magazine feed lips (see fig. 3-8).

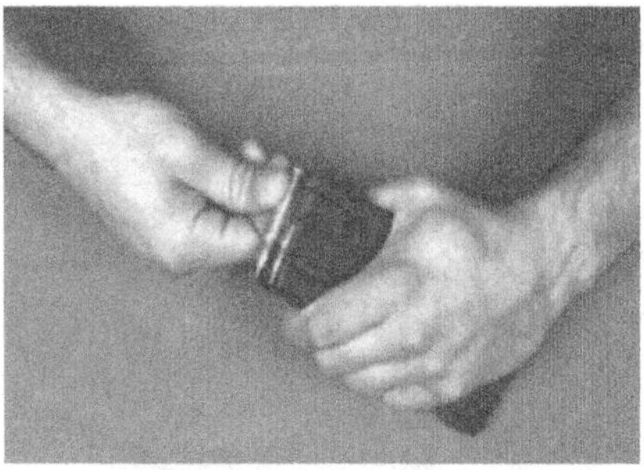

Figure 3-8. Filling the Magazine.

- Repeat until the desired number of rounds is inserted. The recommended number of rounds per magazine is 28 or 29. Thirty rounds in the magazine may prohibit the magazine from seating properly on a closed bolt.
- Tap the back of the magazine to ensure the rounds are seated against the back of the magazine.

Filling the Magazine Using a 10-round Stripper Clip and Magazine Filler

The magazine can also be filled quickly using a 10-round stripper clip and the magazine filler (see fig. 3-9).

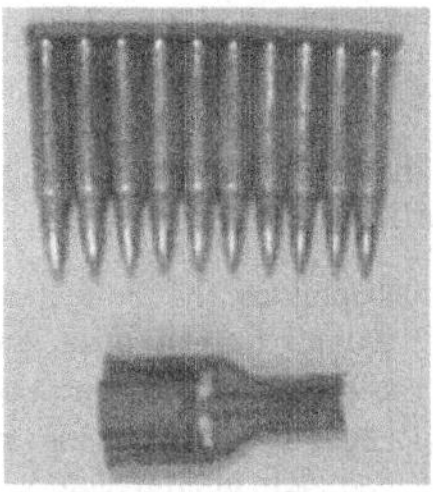

Figure 3-9. Magazine Filler and 10-round

Perform the following steps to fill the magazine with the 10-round stripper clip (see fig. 3-10):

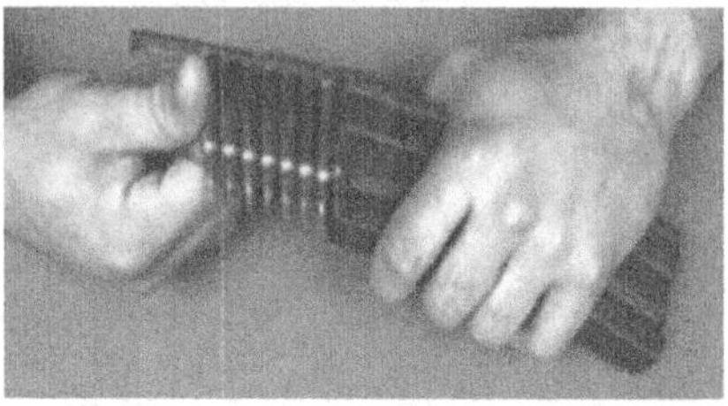

Figure 3-10. Filling the Magazine with a Stripper Clip and Magazine Filler.

- Remove a magazine from the magazine pouch.
- Slide the magazine filler into place.
- Place a 10-round stripper clip into the narrow portion of the magazine filler.
- Using thumb pressure on the rear of the top cartridge, press down firmly until all ten rounds are below the feed lips of the magazine.
- Remove the empty stripper clip while holding the magazine filler in place.
- Repeat until the desired number of rounds is inserted. The recommended number of rounds per magazine is 28 or 29. Thirty rounds in the magazine may prohibit the magazine from seating properly on a closed bolt.
- Remove magazine filler and retain it for future use.
- Tap the back of the magazine to ensure the rounds are seated against the back of the magazine.

Stowing Magazines

Magazine Pouch

In a magazine pouch, filled magazines are stored with rounds down and projectiles pointing away from the body.

Load-bearing Vest

In a load-bearing vest, filled magazines are stored with rounds down and projectiles pointing outboard.

Empty or Partially Filled Magazines

Empty or partially filled magazines are stored with the follower up to allow the selection of filled magazines by touch (i.e., at night).

Withdrawing Magazines

Magazine Pouch

With the right hand, withdraw magazines from the magazine pouch on the right side of the body. With the left hand, withdraw magazines from the magazine pouch on the left side of the body. To withdraw magazines from a magazine pouch—

- Use the thumb and index finger, pinch the magazine pouch release to open the magazine pouch.
- Slide the thumb over the magazines, feeling for a base plate indicating a filled magazine. Continue sliding the thumb until it rests on the back of the magazine.
- Grasp the magazine with the thumb, index finger, and middle finger and lift the magazine directly out of the pouch.
- Once the magazine is clear of the pouch, curl the ring finger and little finger underneath the magazine and rotate it up to observe rounds in the magazine.

Load-bearing Vest

With the right hand, withdraw magazines from the left side of the vest (see fig. 3-11). With the left hand, withdraw magazines from the right side of the vest (see fig. 3-12).

To withdraw magazines from a load-bearing vest:

- With the thumb and index finger, unfasten the snap on the vest pouch.
- Slide the thumb over the magazine, feeling for a base plate indicating a filled magazine.
- Rotate the hand over the magazine while sliding the thumb to the back of the magazine.
- Grasp the magazine with the thumb, index finger, and middle finger, while curling the ring finger and little finger on top of the base plate.

Figure 3-11. Withdrawing a Magazine with the Right Hand.

- Lift the magazine directly out of the pouch and rotate it up to observe the rounds in the magazine.

3006. Reloading the Rifle

Principles of Reloading

The first priority when performing a reload is to get the rifle reloaded and back into action. The second priority when performing a reload is to retain the magazine so when you move, the magazine moves with you. When time permits, retain magazines securely on your person (e.g., in magazine pouch, flak jacket, and cargo pocket). The combat situation may dictate dropping the magazine to the deck when performing a reload. This is acceptable as long as it is picked up before moving to another location.

Take cover before reloading. Always reload before leaving cover to take advantage of the protection provided by cover.

When moving, your focus should be on moving, therefore every effort should be made to not reload while on the move.

When reloading, your focus is on the magazine change.

When reloading, draw the weapon in close to your body so you can see what you are doing and retain positive control of the magazine.

When the new magazine is inserted, tug on it to ensure it is seated. Do not slam the magazine into the weapon hard enough to cause a round to partially pop out of the magazine. This action will cause a double feed and require remedial action.

Figure 3-12. Withdrawing a Magazine

Retain your empty magazines. When there is a lull in the action, refill those magazines so they will be available for future use.

During a lull in the action, replace your magazine when you know you are low on ammunition. This ensures a full magazine of ammunition in the rifle should action resume. Do not wait until the magazine is completely empty to replace it.

Condition 1 Reload

A condition 1 reload is performed when the weapon is in condition 1 by replacing the magazine before it runs out of ammunition. To perform a condition 1 reload, perform the following steps:

- Withdraw a filled magazine from the magazine pouch. With the same hand, press the magazine button and remove the partially filled magazine so it can be retained in the remaining fingers.
- Fully insert the filled magazine into the magazine well and tug downward on the magazine to ensure it is properly seated.
- Store the partially filled magazine in the magazine pouch with rounds up and projectiles pointing away from the body.
- Fasten the magazine pouch.

Dry Reload

A dry reload is required when the magazine in the weapon has been emptied and the bolt has locked to the rear. To perform a dry reload—

- Press the magazine release button.
- Remove the empty magazine and retain it on your person when time permits.
- Fully insert a filled magazine into the magazine well and tug downward on the magazine to ensure it is properly seated.
- Depress the bolt catch to allow the bolt carrier to move forward and observe the round being chambered. This places the rifle in Condition 1.

3007. Remedial Action

If the rifle fails to fire, a Marine performs remedial action. Remedial action is the process of investigating the cause of the stoppage, clearing the stoppage, and returning the weapon to operation.

Observe for Indicators

Once the rifle ceases firing, the Marine must visually or physically observe the ejection port to identify the problem before he can clear it. The steps taken to clear the weapon are based on observation of one of the following three indicators:

Indicator: The bolt is forward or the ejection port cover is closed. See figures 3-13 and 3-14.

Figure 3-13. Bolt Forward.

Figure 3-14. Ejection Port Cover Closed.

To return the weapon to operation—

- Seek cover if the tactical situation permits.
- Tap—Tap the bottom of the magazine.
- Rack—Pull the charging handle to the rear and release it.
- Bang—Sight in and attempt to fire.

Indicator: Brass is obstructing chamber area (usually indicating a double feed or failure to eject). See figure 3-15.

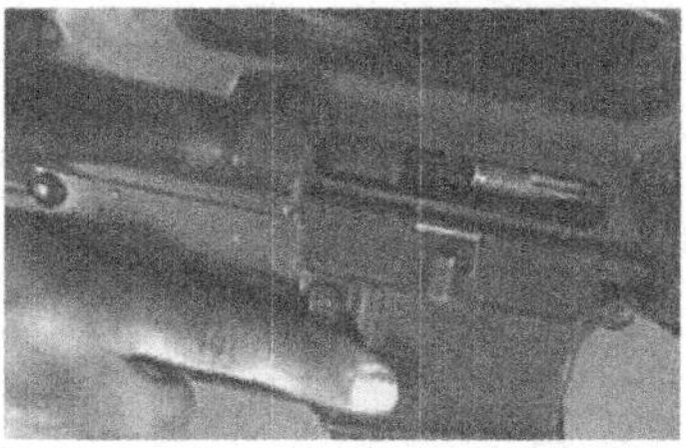

Figure 3-15. Brass Obstructing the Chamber.

To return the weapon to operation—

- Seek cover if the tactical situation permits.
- Attempt to remove the magazine.
- Attempt to lock the bolt to the rear

If the bolt will not lock to the rear, rotate the rifle so the ejection port is facing down; hold the charging handle to the rear as far as it will go and shake the rifle to free the round(s). If the rounds do not shake free, hold the charging handle to the rear and strike the butt of the rifle on the ground or manually clear the round. Conduct a reload. Sight in and attempt to fire.

Indicator: The bolt is locked to the rear. See figure 3-16. To clear return the weapon to operation—

Figure 3-16. Bolt Locked to the Rear.

Note

Although a dry weapon is not considered a true stoppage or mechanical failure, the Marine must take action to return the weapon to operation.

- Seek cover if the tactical situation permits.
- Conduct a dry reload.
- Sight in and attempt to fire.

Audible Pop or Reduced Recoil

An audible pop occurs when only a portion of the propellant is ignited. It is normally identifiable by reduced recoil and is sometimes accompanied by excessive smoke escaping from the chamber area. To clear the rifle in a combat environment:

- Place the rifle in Condition 4.
- Move take down pin from left to right as far as it will go to allow the lower receiver to pivot.
- Remove the bolt carrier group.
- Inspect the bore for an obstruction from the chamber end.
- Insert a cleaning rod into the bore from the muzzle end and clear the obstruction.
- Reassemble the rifle.
- Conduct a reload.
- Sight in and attempt to fire.

3008. Weapons Carries

Weapons carries provide an effective way to handle the rifle while remaining alert to enemy engagement. Weapons carries are tied to threat conditions and are assumed in response to a specific threat situation. The weapons carry assumed prepares the Marine, both mentally and physically, for target engagement. The sling provides additional support for the weapon when firing; therefore, the hasty sling should be used in conjunction with the carries.

Tactical Carry

A Marine carries the rifle at the tactical carry if no immediate threat is present. The tactical carry permits control of the rifle while a Marine is moving, yet it still allows quick engagement of the enemy. A Marine performs the following steps to assume the tactical carry:

- Place left hand on the handguards, right hand around the pistol grip, trigger finger straight along the receiver (see fig. 3-17), and right thumb on top of the selector lever (see fig. 3-18).

Figure 3-17. Straight Trigger Finger.

Figure 3-18. Thumb on Selector Lever.

- Place the buttstock along the side of the body at approximately hip level.
- Angle the muzzle upward about 45 degrees in a safe direction.
- Position the muzzle in front of the eyes, slightly below eye level (see fig. 3-19).
- Move the head and the eyes with the muzzle as it moves.

Figure 3-19. Tactical Carry.

Alert Carry

A Marine carries the rifle at the alert if enemy contact is likely. The alert is also used for moving in close terrain (e.g., urban, jungle). A Marine can engage the enemy faster from the alert than from the tactical carry. However, the alert is more tiring than the tactical carry and its use can be physically demanding. A Marine performs the following steps to assume the alert:

- Place the left hand on the handguards, the right hand around the pistol grip, the trigger finger straight along the receiver (see fig. 3-17), and the right thumb on top of the selector lever (see fig. 3-18).
- Place the buttstock in the shoulder.
- Angle the muzzle downward about 45 degrees and point it in a safe direction or the general direction of likely enemy contact (see fig. 3-20).

Figure 3-20. Alert Carry.

Ready Carry

A Marine carries the rifle at the ready if contact with the enemy is imminent. The ready allows immediate target engagement, but it is very tiring to maintain over a long period of time. A Marine performs the following steps to assume the ready:

- Place left hand on handguards, right hand around the pistol grip, the trigger finger straight along the receiver (see fig. 3-17), and the right thumb on top of the selector lever (see fig. 3-18).
- Place the buttstock in the shoulder.
- Point the muzzle in the direction of the enemy.

- Lower the sights to just below eye level so that a clear field of view is maintained so that a target may be identified (see fig. 3-21).

Figure 3-21. Ready Carry.

3009. Weapons Transports

Weapons transports are used to carry the rifle over the back or shoulders when moving for long periods; they provide a more relaxed position for walking. Weapons transports are used if no immediate threat is present. They are also used whenever one or both hands are needed for other work.

Strong Side Sling Arms Transport (Muzzle Up)

To assume the strong side sling arms (muzzle up) transport from the tactical carry, a Marine performs the following steps (see fig. 3-22):

- Release the hold on the pistol grip.
- Lower the buttstock and bring the rifle to a vertical position.
- With the right hand, grasp the sling above the left forearm.

Figure 3-22. Strong Side Sling Arms Transport (Muzzle Up).

- With the left hand, guide the rifle around the right shoulder.
- With the right hand, apply downward pressure on the sling. This stabilizes the rifle on the shoulder.

Weak Side Sling Arms Transport (Muzzle Down)

The weak side sling arms (muzzle down) transport can be used in inclement weather to keep moisture out of the rifle's bore. To assume this transport from the tactical carry, a Marine performs the following steps (see fig. 3-23):

- Release the hold on the pistol grip.
- With the left hand, rotate muzzle down and bring the rifle to a vertical position on the left side of the body. The pistol grip is pointed outboard.
- With right hand, place sling on left shoulder.
- Grasp sling above the waist with the left hand.
- With the left hand, apply downward pressure on the sling. This stabilizes the rifle on the shoulder.

Figure 3-23. Weak Side Sling Arms Transport (Muzzle Down.)

Cross Body Sling Arms Transport

A Marine uses the cross body sling arms transport if he requires both hands for work. The rifle is slung across the back with the muzzle up or down. Normally, the rifle is carried with the muzzle down to prevent pointing the muzzle in an unsafe direction.

To assume a cross body sling transport, a Marine performs the following steps from weak side sling arms (muzzle down) (see fig. 3-24):

Figure 3-24. Cross Body Sling Arms Transport (Muzzle Down).

- With the right hand, grasp the sling.
- With the left hand, grasp the handguards.
- Pull up on the rifle with both hands.
- Slide the sling over the head.
- Position the rifle so that it rests comfortably across the back.

To assume a cross body sling transport, a Marine performs the following steps from strong side sling arms (muzzle up) (see fig. 3-25):

- With the left hand, grasp the sling.
- With the right hand, grasp the pistol grip.
- Pull up on the rifle with both hands.
- Slide sling over head.
- Position the rifle so that it rests comfortably across the back.

Figure 3-25. Cross Body Sling Arms. Transport (Muzzle Up).

Note

Ensure the muzzle of the rifle is maintained in a safe direction when assuming this transport.

3010. Transferring the Rifle

Proper weapons handling is required every time a Marine picks up a weapon, passes a weapon to another Marine, or receives a weapon from another Marine. It is the responsibility of the Marine receiving or taking charge of a weapon to determine its condition. Depending on the situation, there are two procedures that can be used to transfer a rifle from one Marine to another Marine: show clear transfer and condition unknown transfer.

Show Clear Transfer

When time and the tactical situation permit, the Marine should transfer the rifle using the show clear transfer. To properly pass a rifle between Marines, the Marine handing off the rifle must perform the following procedures:

- Ensure the rifle is on safe.
- Remove the magazine if it is present.
- Lock the bolt to the rear.
- Visually inspect the chamber to ensure there is no ammunition present.
- Leave the bolt locked to the rear and hand the weapon to the other Marine.

The Marine receiving the weapon must:

- Ensure the rifle is on safe.
- Visually inspect the chamber to ensure there is no ammunition present.
- Release the bolt catch and observe the bolt going forward on an empty chamber.
- Close the ejection port cover.

Condition Unknown Transfer

There are times when time or the tactical situation does not permit a show clear transfer of the rifle. The procedures for the condition unknown transfer are conducted by a Marine when he takes charge of a rifle in any situation when the condition of the rifle is unknown (e.g., an unmanned rifle from a casualty, a rifle stored in a rifle rack). To properly take charge of a rifle when its condition is unknown, the Marine must perform the following procedures:

- Ensure the rifle is on safe.
- Conduct a chamber check to determine the condition of the weapon (see para. 3003).
- Remove the magazine and observe if ammunition is present. If time permits, count the rounds.
- Insert the magazine into the magazine well.
- Close the ejection port cover.

3011. Clearing Barrel Procedures

Purpose of a Clearing Barrel

The sole purpose of a clearing barrel is to provide a safe direction in which to point a weapon when loading; unloading; and unloading and showing clear. See figure 3-26 on page 3-12. Clearing barrel procedures are identical to the weapons handling procedures for the rifle for the loading; unloading; and unloading and showing clear.

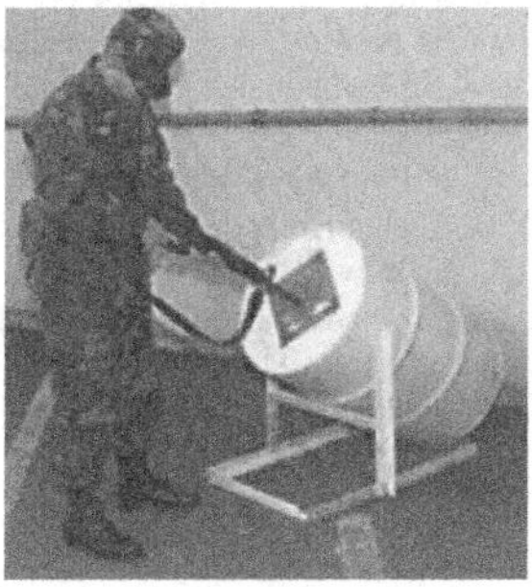

Figure 3-26. Clearing Barrel.

Procedures for "Load"

On the command "Load," the Marine will perform the following steps to take the rifle from Condition 4 to Condition 3:

- With a straight trigger finger, point the rifle in the clearing barrel.
- Ensure the rifle is on safe.
- Withdraw a magazine from the magazine pouch.
- Observe the magazine to ensure it is filled.
- Fully insert the magazine into the magazine well.
- Without releasing the magazine, tug downward on the magazine to ensure it is seated.
- Fasten the magazine pouch.
- Close the ejection port cover.

Procedures for "Make Ready"

If standard operating procedures (SOP) or rules of engagement (ROE) require the rifle to be carried in Condition 1, the Marine will "Make Ready" at the clearing barrel. On the command "Make Ready," the Marine will perform the following steps to take the rifle from Condition 3 to Condition 1:

- Pull the charging handle to the rear and release. There are two methods of doing this:
  - Grip the pistol grip firmly with the right hand and pull the charging handle with the left hand to its rearmost position and release.
  - Or grip the handguards firmly with the left hand and pull the charging handle with the right hand to its rearmost position and release.
- To ensure ammunition has been chambered, conduct a chamber check.
- Check the sights (to ensure proper BZO setting, correct rear sight aperture, etc.).
- Close the ejection port cover.

Procedures for "Unload"

On the command "Unload," the Marine will perform the following steps to take the rifle from any condition to Condition 4:

- With a straight trigger finger, point the rifle in the clearing barrel.
- Ensure the weapon is on safe.
- Remove the magazine from the rifle and retain it on your person.
- While cupping the left hand under the ejection port, rotate weapon until the ejection port is facing down.
- Pull the charging handle to the rear and catch the round in the left hand (see fig. 3-17).
- Lock the bolt to the rear.
- Put weapon on safe if not already on safe.
- Ensure that the chamber is empty and that no ammunition is present.
- Release the charging handle and observe the bolt moving forward on an empty chamber.
- Close the ejection port cover.
- Check the sights (for proper BZO setting, correct rear sight aperture, etc.).
- Place any ejected round into the magazine and return the magazine to the magazine pouch and close the magazine pouch.

Procedures for "Unload and Show Clear"

On the command "Unload and Show Clear," the Marine will perform the following steps to take the rifle from any condition to Condition 4:

- With a straight trigger finger, point the rifle in the clearing barrel.
- Ensure the weapon is on safe.
- Remove the magazine from the rifle and retain it on your person.
- While cupping the left hand under the ejection port, rotate the weapon until the ejection port is facing down.
- Pull the charging handle to the rear and catch the round in the left hand.
- Lock the bolt to the rear.

- Ensure the chamber is empty and no ammunition is present.
- Have an observer inspect the weapon to ensure no ammunition is present. The observer:
  - Visually inspects chamber to ensure it is empty, no ammunition is present, and the magazine is removed.
  - Ensures the weapon is on safe.
  - Acknowledges the rifle is clear.
- After receiving acknowledgment from the observer that the rifle is clear, the Marine releases the bolt catch and observes the bolt moving forward on an empty chamber.
- Close the ejection port cover.
- Check the sights (for proper BZO setting, correct rear sight aperture, etc.).
- Place any ejected round into the magazine and return the magazine to the magazine pouch and close the magazine pouch.

CHAPTER 4. FUNDAMENTALS OF MARKSMANSHIP

The fundamentals of marksmanship are aiming, breathing, and trigger control. These techniques provide the foundation for all marksmanship principles and skills. For rifle fire to be effective, it must be accurate. A rifleman who merely sprays shots in the vicinity of the enemy produces little effect. The fundamentals of marksmanship, when applied correctly, form the basis for delivering accurate fire on enemy targets. These skills must be developed so that they are applied instinctively. During combat, the fundamentals of marksmanship must be applied in a time frame consistent with the size and the distance of the target. At longer ranges, the target appears to be smaller and a more precise shot is required to accurately engage the target. The fundamentals are more critical to accurate engagement as the range to the target increases. To be accurate at longer ranges, the Marine must take the time to slow down and accurately apply the fundamentals. At shorter ranges, the enemy must be engaged quickly before he engages the Marine. As the size of the target increases, and the distance to the target decreases, the fundamentals, while still necessary, become less critical to accuracy.

Note

☞The procedures in this manual are written for right-handed Marines; left-handed Marines should reverse instructions as necessary.

4001. Aiming

Sight Alignment

Sight alignment is the relationship between the front sight post and rear sight aperture and the aiming eye. This relationship is the most critical to aiming and must remain consistent from shot to shot. To achieve correct sight alignment (see fig. 4-1):

- Center the tip of the front sight post vertically and horizontally in the rear sight aperture.
- Imagine a horizontal line drawn through the center of the rear sight aperture. The top of the front sight post will appear to touch this line.
- Imagine a vertical line drawn through the center of the rear sight aperture. The line will appear to bisect the front sight post.

Sight Picture

Sight picture is the placement of the tip of the front sight post in relation to the target while maintaining sight alignment. Correct sight alignment but improper sight placement on the target will cause the bullet to impact the target incorrectly on the spot where the sights were aimed when the bullet exited the muzzle.

To achieve correct sight picture, place the tip of the front sight post at the center of the target while maintaining sight alignment (see fig. 4-2). Center mass is

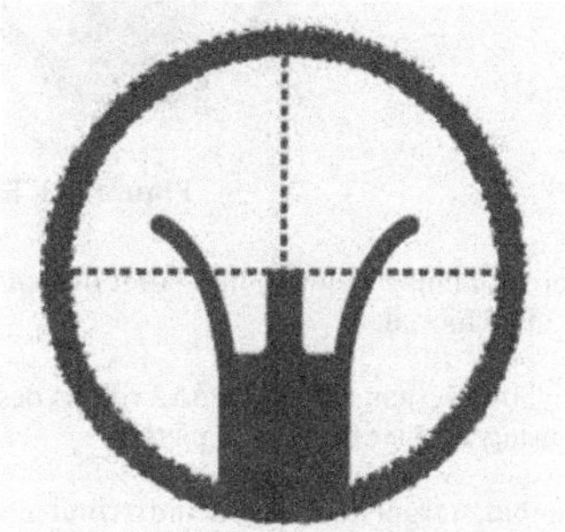

Figure 4-1. Correct Sight Alignment.

Figure 4-2. Correct Sight Picture.

Figure 4-3. Examples of Correct Sight Picture.

the correct aiming point so that point of aim/point of impact is achieved.

The sighting system for the M16A2 rifle is designed to work using a center mass sight picture.

In combat, targets are often indistinct and oddly shaped. The center mass hold provides a consistent aiming point (see fig. 4-3).

Importance of Correct Sight Alignment

A sight alignment error results in a misplaced shot. The error grows proportionately greater as the distance to the target increases. An error in sight picture, however, remains constant regardless of the distance to the target. See figure 4-4.

Factors Affecting Sight Alignment and Sight Picture

Stock Weld

Stock weld is the point of firm contact between the cheek and the stock of the rifle. See figure 4-5. The head should be as erect as possible to enable the aiming eye to look straight through the rear sight aperture. If the position of the Marine's head causes him to look across the bridge of his nose or out from under his eyebrow, the eye will be strained. The eye functions best in its natural forward position. Changing the placement of the cheek up or down on the stock from shot to shot may affect the zero on the rifle due to the perception of the rear sight aperture. A consistent and proper stock weld is critical to the aiming process because it provides consistency in eye relief, which affects the ability to align the sights.

Eye Relief

Eye relief is the distance between the rear sight aperture and the aiming eye. See figure 4-6. Normal eye relief is two to six inches from the rear sight aperture. The distance between the aiming eye and the rear sight aperture depends on the size of the Marine and the firing position. While eye relief varies slightly from one position to another, it is important to have the same eye relief for all shots fired from a particular position.

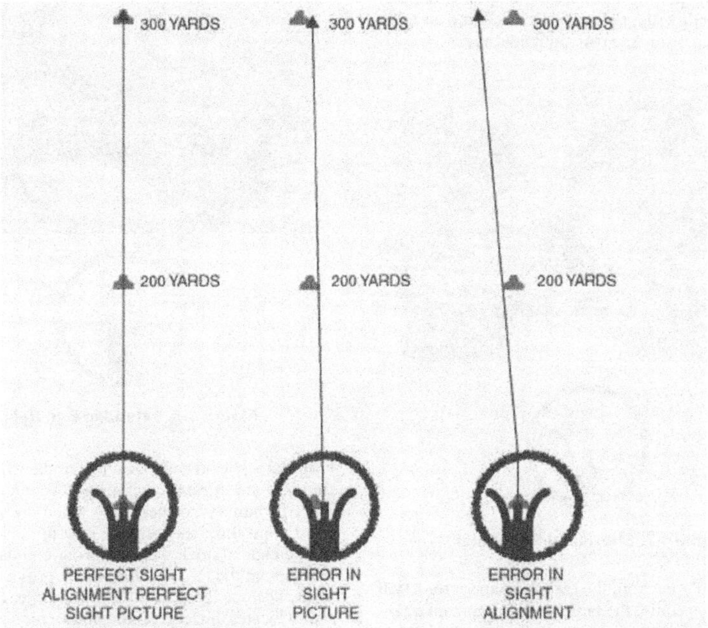

Figure 4-4. Importance of Correct Sight Alignment.

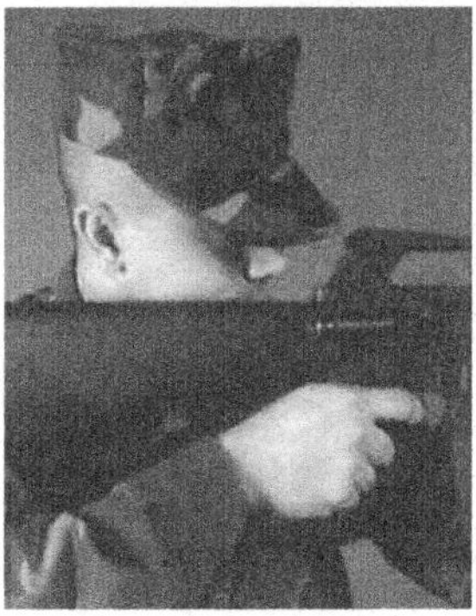

Figure 4-5. Stock Weld.

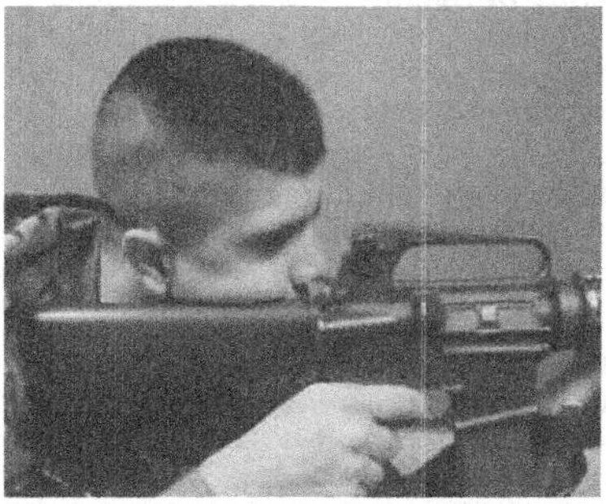

Figure 4-6. Proper Eye Relief.

If the eye is too close to the rear sight aperture, it will be difficult to line up the front sight post in the rear sight aperture (see fig. 4-7). Moving the eye back from the rear sight aperture will make the aperture appear smaller and allow the tip of the front sight post to be easily lined up inside the rear sight aperture.

Figure 4-7. Shortened Eye Relief.

If the eye is too far from the rear sight aperture, it will be difficult to acquire the target and to maintain a precise aiming point (see fig. 4-8).

Wearing of Glasses

Wearing glasses can alter the perception of sight alignment and sight picture. If wearing glasses, it is critical to look through the optic center of the lens.

Acquiring and Maintaining Sight Alignment and Sight Picture

The human eye can focus clearly on only one object at a time. For accurate shooting, it is important to focus on the tip of the front sight post. When the shot is fired, focus must be on the tip of the front sight post; peripheral vision will include the rear sight and the target. The rear sight and the target will appear blurry. Staring or fixing the vision on the front sight post for longer than a few seconds can distort the image, making it difficult to detect minute errors in sight alignment.

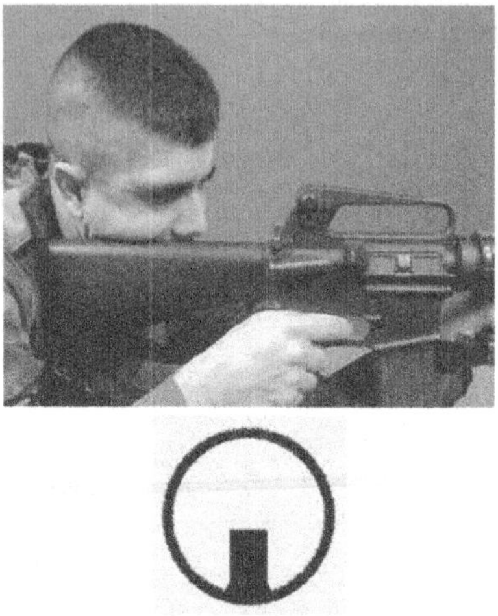

Figure 4-8. Extended Eye Relief.

Proper stock weld and placement of the rifle butt in the shoulder aid in establishing sight alignment quickly. The rifle butt's placement in the shoulder serves as the pivot point for presenting the rifle up to a fixed point on the cheek (stock weld). During combat, a Marine will look at the target as the rifle is presented. As rifle sights become level with the aiming eye, a Marine visually locates the target through the rear sight aperture. As the rifle settles, a Marine's focus shifts back to the front sight post to place the tip of the post on the target and obtain sight alignment and sight picture. To maintain sight alignment and sight picture, the Marine's focus should shift repeatedly from the front sight post to the target until correct sight alignment and sight picture are obtained. This enables the detection of minute errors in sight alignment and sight picture.

Size and Distance to the Target

During combat, the fundamentals of marksmanship must be applied in a time frame consistent with the size and the distance to the target.

Long-range Engagements

At longer ranges, the target appears smaller and a more precise shot is required to accurately engage the target. Sight alignment and sight picture are more critical to accurate engagement as the range to the target increases. To be accurate at longer ranges, the Marine must take the time to slow down and accurately apply the fundamentals.

As the distance to the target increases, the front sight post covers more of the target, making it difficult to establish a center of mass hold (see fig. 4-9). Since the Marine must see the target to engage it, there is a tendency to look at the target by lowering the tip of the front sight post. This causes shots to impact low or miss the target completely. A Marine must consciously aim at the center of mass and attempt to maintain a center mass sight picture.

Short-range Engagements

At shorter ranges, the enemy must be engaged quickly before he engages the Marine. As distance to the target decreases, the size of the target appears to increase, and sight alignment becomes less critical to accuracy. At very short ranges, a deviation in sight alignment can still produce accurate results as long as the tip of the front sight post is in the rear sight aperture and on the target (see fig. 4-10). The time required to engage a target is unique to each individual. Although a Marine must engage the target rapidly, some semblance of sight alignment is still required to be accurate.

4002. Breath Control

Proper breath control is critical to the aiming process. Breathing causes the body to move. This movement transfers to the rifle making it impossible to maintain proper sight picture. Breath control allows the Marine to fire the rifle at the moment of least movement.

Breath Control During Long-range or Precision Fire (Slow Fire)

It is critical that Marines interrupt their breathing at a point of natural respiratory pause before firing a long-range shot or a precision shot from any distance. A respiratory cycle lasts 4 to 5 seconds. Inhaling and exhaling each require about 2 seconds. A natural pause of 2 to 3 seconds occurs between each respiratory cycle. The pause can be extended up to 10 seconds. During the pause, breathing muscles are relaxed and the sights settle at their natural point of aim. To minimize movement, Marines must fire the shot during the natural respiratory pause. The basic technique is as follows:

- Breathe naturally until the sight picture begins to settle.
- Take a slightly deeper breath.
- Exhale and stop at the natural respiratory pause.
- Fire the shot during the natural respiratory pause.

Note

If the sight picture does not sufficiently settle to allow the shot to be fired, resume normal breathing and repeat the process.

Breath Control During All Other Combat Situations

A Marine in a combat environment may not have the time to fire a shot during the natural respiratory pause.

Figure 4-9. Sight Picture at Long-range Engagements.

Figure 4-10. Sight Picture at Short-range Engagements.

It may be necessary to take several deep breaths before holding the breath. A Marine should not make an exaggerated effort to perform breath control. A natural respiratory pause will help stabilize the shooter's sight picture. The basic technique is as follows:

- Take a deep breath filling the lungs with oxygen.

Note

It may be necessary to take several deep breaths quickly before holding the breath.

- Hold the breath and apply pressure to the trigger.
- Fire the shots.

4003. Trigger Control

Trigger control is the skillful manipulation of the trigger that causes the rifle to fire without disturbing sight alignment or sight picture. Controlling the trigger is a mental process, while pulling the trigger is a physical process.

Grip

A firm grip is essential for effective trigger control. The grip is established before starting the application of trigger control and it is maintained through the duration of the shot. To establish a firm grip on the rifle, position the "V" formed between the thumb and index finger on the pistol grip behind the trigger. The fingers and the thumb are placed around the pistol grip in a location that allows the trigger finger to be placed naturally on the trigger and the thumb in a position to operate the safety. Once established, the grip should be firm enough to allow manipulation of the trigger straight to the rear without disturbing the sights. See figure 4-11.

Trigger Finger Placement

Correct trigger finger placement allows the trigger to be pulled straight to the rear without disturbing sight alignment. The trigger finger should contact the trigger naturally. The trigger finger should not contact the rifle receiver or trigger guard.

Figure 4-11. Grip.

Types of Trigger Control

There are two techniques of trigger control: uninterrupted and interrupted

Uninterrupted Trigger Control

The preferred method of trigger control in a combat environment is uninterrupted trigger control. After obtaining sight picture, the Marine applies smooth, continuous pressure rearward on the trigger until the shot is fired.

Interrupted Trigger Control

Interrupted trigger control is used at any time the sight alignment is interrupted or the target is temporarily obscured. An example of this is extremely windy conditions when the weapon will not settle, forcing the Marine to pause until the sights return to his aiming point. To perform interrupted trigger control:

- Move the trigger to the rear until an error is detected in the aiming process.
- When this occurs, stop the rearward motion on the trigger, but maintain the pressure on the trigger, until sight picture is achieved.
- When the sight picture settles, continue the rearward motion on the trigger until the shot is fired.

Resetting the Trigger

During recovery, release the pressure on the trigger slightly to reset the trigger after the first shot is delivered (indicated by an audible click). Do not remove the finger from the trigger. This places the trigger in position to fire the next shot without having to reestablish trigger finger placement.

4004. Follow-Through/Recovery

Follow-Through

Follow-through is the continued application of the fundamentals until the round has exited the barrel. In combat, follow-through is important to avoid altering the impact of the round by keeping the rifle as still as possible until the round exits the barrel.

Recovery

It is important to get the rifle sights back on the target for another shot. This is known as recovery. Shot recovery starts immediately after the round leaves the barrel. To recover quickly, a Marine must physically bring the sights back on target as quickly as possible.

CHAPTER 5. RIFLE FIRING POSITIONS

In a combat environment, a Marine must be prepared to engage the enemy under any circumstance. There are four basic firing positions: prone, sitting, kneeling, and standing. These positions provide a stable foundation for effective shooting. Any firing position must provide stability, mobility, and observation of the enemy. During training, a Marine learns positions in a step-by-step process, guided by a series of precise movements until the Marine assumes a correct position. The purpose of this is to ensure that the Marine correctly applies all of the factors that assist him in holding the rifle steady. The Marine will gradually become accustomed to the feel of the positions through practice and eventually will be able to know instinctively whether his position is correct. In combat, it may not be possible to assume a textbook firing position due to terrain, available cover, engagement time, dispersion of targets, and other limiting factors. Modifications to the basic positions may have to be made to adjust to the combat environment. The Marine must strive to assume a position that offers stability for firing, maximum cover and concealment from the enemy, and maximum observation of the target.

Note

+The procedures in this manual are written for right-handed Marines; left-handed Marines should reverse instructions as necessary.

5001. Selecting a Firing Position

The selection of a firing position (prone, sitting, kneeling, standing) is based on terrain, available cover, dispersion of targets, and other limiting factors. A Marine must select a position that offers stability, mobility, and observation.

Stability

A firing position must provide a stable platform for accurate and consistent shooting. If the position is solid, the front sight can be held steady and the rifle sights should recover after recoil to the same position on the target. This allows for rapid reengagement of the enemy. The prone position provides the most stability for firing, while the standing position provides the least stability.

Mobility

A firing position must provide a Marine with the mobility required to move to new cover or to another area. The standing position permits maximum mobility, and it also allows the most lateral movement for engagement of widely dispersed targets. The prone position allows the least mobility and allows limited lateral movement.

Observation of the Enemy

A firing position must limit a Marine's exposure to the enemy, yet allow observation of the enemy. Manmade structures and terrain features (e.g., vegetation, earth contours) often dictate the shooting position. The standing position normally provides the best field of view, but it usually allows the most exposure to the enemy. The prone position normally allows the least exposure, but it usually provides a limited field of view.

5002. Types and Uses of the Rifle Web Sling

The rifle sling, when adjusted properly, provides maximum stability for the weapon, and helps hold the front sight still and reduce the effects of the rifle's recoil. Once a sling adjustment is found that provides maximum control of the weapon, the same sling adjustment should be maintained. Varying the sling tension extensively will affect the strike of the bullet, which will make maintaining a BZO difficult. Using the same sling adjustment will ensure the accuracy of rounds on target. There are two basic types of rifle sling adjustments: the hasty sling and loop sling.

Note

In training situations, the parade sling may be used. See paragraph 5008.

Hasty Sling

Application

The hasty sling is used in all firing positions. The hasty sling is advantageous in combat because it can be acquired quickly and it provides added stability to the rifle. The same sling setting can be used for all firing positions. If properly adjusted, the hasty sling supports the weight of the weapon, provides stability for the rifle, and reduces the effects of the rifle's recoil. When using the hasty sling, controlled muscle tension is applied to offer resistance against the sling, enabling the rifle sights to be held steady.

Donning the Hasty Sling

A Marine performs the following steps to form a hasty sling:

- Hold the rifle vertical with the barrel pointing upward.
- Unhook the J-hook from the lower sling swivel.
- Loosen the sling keeper.
- Adjust the sling until the J-hook hangs below the rifle butt. (The distance will vary based on the individual Marine, but the J-hook will usually hang approximately 3 to 10 inches below the rifle butt.) Secure the sling keeper. See figure 5-1.

Figure 5-1. J-hook Location.

- Turn the sling a half turn outboard to allow the sling to lay flat against the arm.
- Attach the J-hook to the lower sling swivel so the open end of the J-hook faces outboard, away from the rifle. See figure 5-2.

Figure 5-2. J-hook Turned Outboard.

- While holding the rifle with the right hand, place the left arm through the sling near the lower sling swivel. Slide the arm up through the sling below the half twist. The sling makes contact low on the arm just below the triceps, above the elbow. The sling lies flat on the back of the arm.
- With the left hand, grasp the handguard by pinching it in the "V" formed by the thumb and forefinger. The sling lies flat against the back or side of the wrist or on the arm near the wrist. See figure 5-3.

Figure 5-3. Sling Against the Wrist.

- Move the left hand as required to level the rifle with the line of sight. Placement of the forward hand controls the tension on the sling between the back of the wrist or arm and the upper sling swivel (see fig. 5-4). This hand placement, with a straight locked wrist, will cause the sling to pull straight under the handguards and serves to stabilize the front sight of the rifle.

- Move the feed end of the sling in or out of the sling keeper to adjust the hasty sling. Sling tension is in-

Figure 5-4. Position of Forward Hand.

creased by pushing the elbow outboard. See figure 5-5. (This enables one sling setting to fit all positions.) It is important for the hasty sling to be adjusted so it supports the rifle. The sling setting must allow the left elbow to push outboard against the sling so the elbow is not inverted under the rifle.

Figure 5-5. Position of Left Elbow.

- Locate the sling keeper near the feed end of the sling and secure so the backside or flat end of the sling keeper is against the arm.

Loop Sling

Application

The loop sling provides the greatest amount of stability during firing. This stability allows the Marine to perfect marksmanship fundamentals. A loop sling takes longer to don or remove than a hasty sling. Therefore, it has limited combat application and is best used where stability of hold is needed for a precision or long-range shot. The loop sling is used in the prone, sitting, and kneeling positions.

Donning the Loop Sling

- Place the rifle butt on the right hip and cradle the rifle in the right arm.
- Disconnect the J-hook from the lower sling swivel.
- With the M-buckle near the hook, feed the sling through the top of the M-buckle to form a loop large enough to slip over the arm. See figure 5-6.
- Give the loop a half turn outboard and insert the left arm through the loop, positioning the loop above the biceps. The loop is high on the left arm above the biceps muscle in such a position that it does not transmit pulse beat to the rifle.
- Position the M-buckle on the outside of the left arm. See figure 5-7.
- Tighten the loop on the left arm, ensuring the M-buckle moves toward the center of the arm as the loop tightens. The sling must pull from the center of the arm to be properly positioned. In this way, as tension is applied to the sling in the firing position, the loop will tighten.

To adjust the sling for the proper length, loosen the sling keeper and pull up or down (toward or away) from the loop. This adjustment varies with every individual and every firing position. The loop should not be tightened excessively on the arm. If blood flow is restricted, excessive pulse beat is transmitted through the rifle sling to the rifle and causes a noticeable, rhythmic movement of the rifle sights. When this occurs, a stable hold at the desired aiming point is impossible to achieve.

Figure 5-6. Forming a Loop.

Figure 5-7. Position of M-buckle.

Tension on the rifle sling is correct when it causes the rifle butt to be forced rearward into the pocket of the shoulder. This serves to keep the butt plate in the shoulder pocket during recoil. To increase tension on the rifle sling, the sling must be shortened. To lessen the tension, the rifle sling must be lengthened.

Move the sling keeper toward the left arm and secure it. The sling keeper should be positioned near the feed end of the sling.

Place the left hand over the sling from the left side and under the rifle. The rifle handguard should rest in the "V" formed between the thumb and forefinger and across the palm of the hand.

Move the left hand as required to achieve desired sight picture. Adjust the length of the sling for proper sling tension and support. See figure 5-8.

Figure 5-8. Loop Sling Donned.

5003. Factors Common to All Shooting Positions

There are seven factors that are common to all shooting positions that affect the ability to hold the rifle steady, maintain sight alignment, and control the trigger. The way these factors are applied differs slightly for each position, but the principles of each factor remain the same.

Left Hand

Placement of the left hand affects placement of left elbow, eye relief, stock weld, and sling tension.

Hasty Sling

In a hasty sling configuration, the sling is attached to the upper and lower sling swivels of the rifle. When the left arm is placed in the hasty sling, tension created by the sling travels from side to side. The tension created by the sling affects how the position is established. There are fundamental differences between the application of the seven factors when using the hasty sling. The most obvious of these is placement of the left hand and the left elbow.

To maximize the support provided by the hasty sling, the left elbow should not be inverted and under the rifle. Instead, the elbow should be pushed outboard against the sling. To achieve this, the position of the shooter's body must almost face the target as opposed to being perpendicular to the target. In addition, the hasty sling must be loosened to allow the elbow to push out against the sling far enough so that the elbow is not under the rifle.

The tension on the sling created by the hasty sling causes the center of balance to change on the rifle. When the elbow is under the rifle with the hasty sling donned, the sling pulls down on the sling swivel disrupting the center of balance and causing the muzzle to drop. Therefore, the elbow must be pushed outboard.

Outboard tension on the sling by the elbow drives the buttstock into the pocket of the shoulder. To enable this, the sling must make contact on the arm just below the triceps, above the elbow. See figure 5-9.

Figure 5-9. Position of Left Elbow with Hasty Sling.

To stabilize the front sight of the rifle, the forward hand, wrist, and forearm should be straight with the wrist locked in place; the hand should be rotated up so the rifle rests in the "V" formed by the thumb and index finger; the fingers will not curl around the handguards. Instead, they will pinch the handguard slightly to keep the hand from slipping on the handguard during recoil. See figure 5-10.

Figure 5-10. Position of Left Hand with Hasty Sling.

When the wrist of the left hand is straight and locked, it creates resistance on the sling close to the muzzle. The sling is in contact with the back or side of the wrist or on the arm near the wrist. This resistance allows the front sight to be stabilized.

In contrast, when the rifle rests across the palm of the hand, the only resistance created is where the sling meets the triceps. Since the resistance is further from the muzzle of the rifle, it makes stabilizing the front sight more difficult.

Loop Sling

With the loop sling donned, the handguard of the rifle rests in the "V" formed by the thumb and index finger of the left hand. The rifle rests across the heel of the hand. The left elbow should be positioned under the weapon to create bone support and a consistent resistance to recoil. The fingers can curl around the handguard, but should apply only the minimum amount of pressure to prevent the hand from slipping on the handguard. See figure 5-11.

Figure 5-11. Position of Left Hand with Loop Sling.

Rifle Butt in the Pocket of the Shoulder

The rifle butt placed firmly in the pocket formed in the right shoulder provides resistance to recoil, helps steady the rifle, and prevents the rifle butt from slipping in the shoulder during firing. Consistent placement of the rifle butt in the shoulder pocket is essential to maintaining a BZO and firing tight shot groups.

Hasty Sling

With the hasty sling donned, the butt of the rifle is placed in the pocket of the shoulder. The body is squared to the target to provide a pocket for the butt of the weapon.

Outboard tension on the sling by the left elbow drives the buttstock into the pocket of the shoulder.

The rifle butt should be placed high in the shoulder to achieve a proper stock weld, allowing the Marine to bring the stock up to his head, rather than lower his head to the stock, which can degrade acquisition of sight alignment and sight picture. See figure 5-12.

Figure 5-12. Buttstock in the Shoulder with Hasty Sling.

Loop Sling

With the loop sling donned, the toe of the rifle butt is placed in the pocket of the shoulder. See figure 5-13.

Figure 5-13. Buttstock in the Shoulder with Loop Sling.

Grip of the Right Hand

Grasp the pistol grip with the right hand and place the forefinger on the trigger, with the thumb and remaining fingers wrapped around the pistol grip. See figure 5-14.

Figure 5-14. Grip of the Right Hand.

Firm rearward pressure should be exerted to help keep the rifle butt firmly in the shoulder, reducing the effects of recoil (this type of pressure is particularly true with the loop sling).

The trigger finger should be placed naturally on the trigger and care should be taken to ensure that the trigger finger can move independently without dragging on the side of the receiver. Proper placement of the right hand high on the pistol grip allows the trigger to be moved straight to the rear without disturbing sight alignment.

Right Elbow

The right elbow should be positioned naturally to provide balance to the position and create a pocket in the shoulder for the rifle butt. If the elbow is correctly positioned, it helps to form a pocket in the right shoulder where the rifle butt rests. The exact placement of the elbow varies with each shooting position but should remain consistent from shot to shot, ensuring the resistance to recoil remains constant. See figure 5-15.

Figure 5-15. Right Elbow.

Stock Weld

The placement of the shooter's cheek against the stock should remain firm and consistent from shot to shot. Consistency of stock weld is achieved through proper placement of the rifle butt in the pocket of the shoulder. A firm contact between the cheek and the stock enables consistent eye relief and enables the head and rifle to recoil as a single unit. Stock weld provides quick recovery between rapid fire shots, keeps the aiming eye centered in the rear sight aperture, and prevents the head from bouncing off the stock during recoil. The head should remain erect to allow the aiming eye to look straight through the rear sight aperture. See figure 5-16.

Figure 5-16. Stock Weld.

Breathing

Breathing causes chest movement and a corresponding movement in the rifle and its sights. Applying breath control will minimize this movement and the effect it has on aiming.

Muscular Tension/Relaxation

Muscular Tension–Hasty Sling

With the hasty sling donned, the shooter must apply an amount of controlled muscular tension in the left arm to keep the sling taut and stabilize the weapon sights. Resistance against the hasty sling controls the point at which the rifle sights will settle. The muscular tension is applied outward against the sling rather than in an effort to hold the rifle up. However, muscular tension should not be excessive to cause the shooter to shake, tremble, or experience fatigue.

Muscular Relaxation–Loop Sling

When using the loop sling, the muscles should be relaxed. Relaxation prevents undue muscle strain and reduces excessive movement. If proper relaxation is achieved, natural point of aim and sight alignment are more easily maintained.

5004. Elements of a Good Shooting Position

There are three elements of a good shooting position that apply when using a loop sling: bone support, muscular relaxation, and natural point of aim. The three elements of a shooting position applied with the loop sling do not apply in the same way as when firing with a hasty sling. While some degree of bone support is still achieved with the hasty sling, muscular **tension** is applied rather than muscular **relaxation**. Natural point of aim, however, applies to both the loop sling and the hasty sling.

Bone Support

The body's skeletal structure provides a stable foundation to support the rifle's weight. A weak shooting position will not withstand a rifle's repeated recoil when firing at the sustained rate or buffeting from wind. To attain a correct shooting position, the body's bones must support as much of the rifle's weight as possible. Proper use of the sling provides additional support.

The weight of the weapon should be supported by bone rather than muscle because muscles fatigue whereas bones do not.

By establishing a strong foundation for the rifle utilizing bone support, the Marine can relax as much as possible while minimizing weapon movement due to muscle tension.

Muscular Relaxation

Once bone support is achieved, muscles are relaxed. Muscular relaxation helps to hold the rifle steady and increase the accuracy of the aim. Muscular relaxation also permits the use of maximum bone support to create a minimum arc of movement and consistency in resistance to recoil. Muscular relaxation cannot be achieved without bone support. During the shooting process, the muscles of the body must be relaxed as much as possible. Muscles that are tense will cause excessive movement of the rifle, disturbing the aim. When proper bone support and muscular relaxation are achieved, the rifle will settle onto the aiming point, making it possible to apply trigger control and deliver a well-aimed shot.

Natural Point of Aim

The point at which the rifle sights settle when in a firing position is called the natural point of aim.

Since the rifle becomes an extension of the body, it may be necessary to adjust the position of the body until the rifle sights settle naturally on the desired aiming point on the target.

When in a shooting position with proper sight alignment, the position of the tip of the front sight post will indicate the natural point of aim. When completely relaxed, the tip of the front sight post should rest on the desired aiming point.

One method of checking for natural point of aim is to aim in on the target, close the eyes, take a couple of breaths, and relax as much as possible. When the eyes are opened, the tip of the front sight post should be positioned on the desired aiming point while maintaining sight alignment.

For each shooting position, specific adjustments will cause the rifle sights to settle center mass, achieving a natural point of aim.

In all positions, the natural point of aim can be adjusted by—

- Varying the placement of the left hand in relation to the handguards.
- Moving the left hand forward on the handguards to lower the muzzle of the weapon, causing the sights to settle lower on the target.
- Moving the left hand back on the handguards to raise the muzzle of the weapon, causing the sights to settle higher on the target.
- Varying the placement of the stock in the shoulder.
- Moving the stock higher in the shoulder to lower the muzzle of the weapon, causing the sights to settle lower on the target.
- Moving the stock lower in the shoulder to raise the muzzle of the weapon, causing the sights to settle higher on the target.
- Natural point of aim can be adjusted right or left by adjusting body alignment in relation to the target.

In the prone position, if the natural point of aim is above or below the desired aiming point, move the body slightly forward or back using the left elbow as a pivot and by digging the toes in.

- Pushing the body forward causes the sights to settle lower on the target.
- Pulling the body backward causes the sights to settle higher on the target.

In the kneeling and sitting positions, natural point of aim can be adjusted by varying the placement of the left elbow on the knee.

- Moving the left elbow forward on the knee lowers the muzzle of the weapon, causing the sights to settle lower on the target.
- Moving the left elbow back on the knee raises the muzzle of the weapon, causing the sights to settle higher on the target.

5005. Prone Position

Application

The prone position provides a very steady foundation for shooting and presents a low profile for maximum concealment. However, the prone position is the least mobile of the shooting positions and may restrict a Marine's field of view for observation. In this position, the Marine's weight is evenly distributed on the elbows, providing maximum support and good stability for the rifle.

Assuming the Prone Position

The prone position can be assumed by either moving forward or dropping backward into position, depending on the combat situation.

Moving Forward into Position

To move forward into the prone position, the Marine performs the following steps:

- With the hasty sling donned, stand erect and face the target, keeping the feet a comfortable distance apart (approximately shoulder width).
- Place the left hand on the handguard, the right hand on the pistol grip.
- Lower yourself into position by dropping to both knees (see fig. 5-17).

Figure 5-17. Dropping to Both Knees.

- Then shift the weight forward to lower the upper body to the ground using the right hand to break the forward motion. See figure 5-18.

Figure 5-18. Moving Forward Into Position.

Dropping Back into Position

It may be necessary to drop backward into position to avoid crowding cover, or to avoid covering uncleared terrain. To drop back into the prone position, the Marine performs the following steps:

- Face the target.
- Place the left hand on the handguard, the right hand on the pistol grip.
- Squat to the ground and break the fall with either hand (see fig. 5-19).
- Kick both legs straight to the rear (see fig. 5-20).
- If the fall was broken using the left hand, reestablish the hasty sling.

Figure 5-19. Breaking the Fall.

Figure 5-20. Kicking Back Into Position.

Straight Leg Prone Position with the Hasty Sling

Apply the seven factors to this position (para. 5003). To assume the straight leg prone position with the hasty sling, either move forward or drop back into position (see figs. 5-21, 5-22, and 5-23):

- Once on the ground, extend your left elbow in front of you. Stretch your legs out behind you. Spread the feet a comfortable distance apart with the toes pointing outboard and the inner portion of the feet in contact with the ground.
- As much of the body mass should be aligned directly behind the rifle as possible.
- If body alignment is correct, weapon recoil is absorbed by the whole body and not just the shoulder.
- Grasp the pistol grip with the right hand and place the rifle butt in the right shoulder pocket.
- Lower the head and place the cheek firmly against the stock to allow the aiming eye to look through the rear sight aperture.
- Rotate the left hand up, pinching the handguard between the thumb and forefinger.
- Slide both elbows outboard on the ground so there is outboard tension against the sling (moving the elbows out tightens the sling) and both shoulders are level. The elbows should provide a tripod of support with the body.
- Adjust the position of the left hand on the handguard to allow the sling to support the weapon and the front sight to be centered in the rear sight aperture. To adjust for elevation:
  - Move the left hand rearward or forward on the handguards (moving the hand rearward elevates the muzzle).
  - Open or close "V" of the left hand for small adjustments. (closing the "V" elevate the muzzle).
- To adjust for a minor cant in the rifle, rotate the handguard left or right in the "V" formed by the thumb and forefinger by rotating the pistol grip left or right.

Straight Leg Prone Position with the Hasty Sling

Figure 5-21. Front View.

Figure 5-22. Left View.

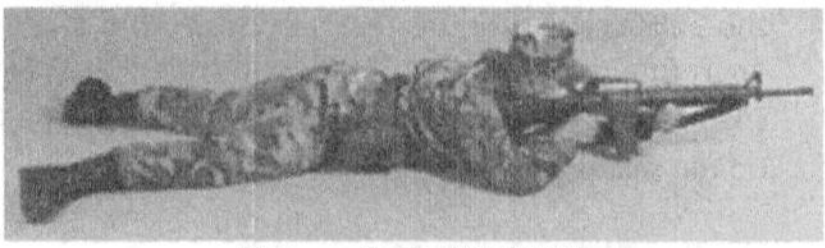

Figure 5-23. Right View.

Straight Leg Prone Position with the Loop Sling

Figure 5-24. Front View.

Figure 5-25. Left View.

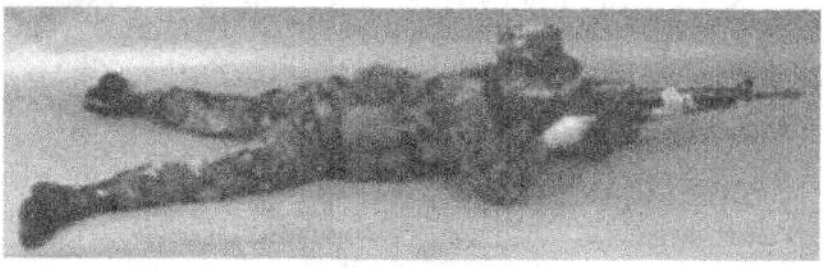

Figure 5-26. Right View.

Straight Leg Prone Position with the Loop Sling

Apply the three elements and seven factors to this position (para. 5004). To assume straight leg prone position with the loop sling, either move forward or drop back into position (see figs. 5-24, 5-25, and 5-26):

- Once on the ground, roll the body to the left side as you extend and invert the left elbow on the ground. Stretch your legs out behind you. Spread the feet a comfortable distance apart with the toes pointing outboard and the inner portion of the feet in contact with the ground.
- As much of the body mass should be aligned directly behind the rifle as possible.
- If body alignment is correct, weapon recoil is absorbed by the whole body and not just the shoulder.
- Grasp the rifle butt with the right hand and place the rifle butt into the right shoulder pocket.
- Grasp the pistol grip with the right hand.
- Rotate the body to the right while the elbow is lowered to the ground so the shoulders are level. The right hand pulls and holds the rifle in the shoulder.
- Lower the head and place the cheek firmly against the stock to allow the aiming eye to look through the rear sight aperture.
- Move the left hand to a location under the handguard, which provides maximum bone support and stability for the weapon. This may require that you remove the rifle from the shoulder to reposition the left hand.

Cocked Leg Prone Position with the Hasty Sling

Figure 5-27. Front View.

Figure 5-28. Left View.

Figure 5-29. Right View.

Cocked Leg Prone Position with the Hasty Sling

Apply the seven factors to this position (para. 5003). To assume the cocked leg prone position with the hasty sling, either move forward or drop back into position (see figs. 5-27, 5-28, and 5-29):

- Once on the ground, roll the body to the left side. The left leg is stretched out behind you, almost in a straight line. This allows the mass of the body to be placed behind the rifle to aid in absorbing recoil.
- Turn the toe of the left foot inboard so the outside of the left leg and foot are in contact with the ground. Bend the right leg and draw it up toward the body to a comfortable position. Turn the right leg and foot outboard so the inside of the right boot is in contact with the ground. Cocking the leg will raise the diaphragm, making breathing easier.
- Grasp the pistol grip with the right hand and place the rifle butt in the right shoulder pocket.
- Lower the head and place the cheek firmly against the stock to allow the aiming eye to look through the rear sight aperture.
- Rotate the left hand up, pinching the handguard between the thumb and forefinger.
- Roll the body to the right while lowering the right elbow to the ground. Slide both elbows outboard on the ground so there is outboard tension against the sling (moving the elbows out tightens the sling). The right shoulder is higher than the left shoulder in the cocked leg position.
- Adjust the position of the left hand on the handguard to allow the sling to support the weapon and the front sight to be centered in the rear sight aperture. To adjust for elevation—
  - Move the left hand rearward or forward on the handguards (moving the hand rearward elevates the muzzle).
  - Open or close "V" of the left hand for small adjustments (closing the "V" elevates the muzzle).

To adjust for a minor cant in the rifle, rotate the handguard left or right in the "V" formed by the thumb and forefinger by rotating the pistol grip left or right.

Cocked Leg Prone Position with the Loop Sling

Apply the three elements and seven factors to this position (para. 5004). To assume the cocked leg prone position with the loop sling, either move forward or drop back into position (see figs 5-30, 5-31, and 5-32):

- Once on the ground, roll the body to the left side and extend and invert the left elbow on the ground. The left leg is stretched out behind you, almost in a straight line. This allows the mass of the body to be placed behind the rifle to aid in absorbing recoil.
- Turn the toe of the left foot inboard so the outside of the left leg and foot are in contact with the ground. Then bend the right leg and draw it up toward the body to a comfortable position. Turn the right leg and foot outboard so the inside of the right boot is in contact with the ground. Cocking the leg will raise the diaphragm, making breathing easier.
- Grasp the rifle butt with the right hand and place the rifle butt into the right shoulder pocket.

Cocked Leg Prone Position with the Loop Sling

Figure 5-30. Front View.

Figure 5-31. Left View.

Figure 5-32. Right View.

- Grasp the pistol grip with the right hand.
- Roll the body to the right while lowering the right elbow to the ground. The right shoulder is higher than the left shoulder in the cocked leg position.
- Lower the head and place the cheek firmly against the stock to allow the aiming eye to look through the rear sight aperture.
- Move the left hand to a location under the handguard which provides maximum bone support and stability for the weapon.

5006. Sitting Position

There are three variations of the sitting position: crossed ankle, crossed leg, and open leg. Experiment with all the variations and select the position that provides the most stability for firing.

Although the sitting position provides an extremely stable base, it limits lateral movement and maneuver ability. It has several variations that can be adapted to the individual Marine. The sitting position provides greater elevation than the prone position while still having a fairly low profile.

Crossed Ankle Sitting Position with the Hasty Sling

The crossed ankle sitting position is an extremely stable shooting position. This position places most of the body's weight behind the weapon and aids in quick shot recovery. Apply the seven factors to this position (para. 5003). To assume the crossed ankle sitting position with the hasty sling (see figs. 5-33, 5-34, 5-35)—

- Square the body to the target.
- Grasp the handguard with the left hand.
- Bend at knees and break the fall with the right hand.
- Push backward with the feet to extend the legs and place the buttocks on the ground.
- Cross the left ankle over the right ankle.
- Grasp the pistol grip with the right hand and place the rifle butt in the right shoulder pocket.
- Lower the head and place the cheek firmly against the stock to allow the aiming eye to look through the rear sight aperture.
- Rotate the left hand up, pinching the handguard between the thumb and forefinger.
- Bend forward at the waist and place the flat portion of the back of the left arm, just above the elbow, on the left leg just below the knee.
- Place the right elbow on the inside of the right knee.
- Adjust the position to adjust sling tension. To tighten the sling: square the body more to the target, or move the left elbow out, or draw the feet up slightly toward the body.
- Adjust the position of the left hand to allow the sling to support the weapon and the front sight to be centered in the rear sight aperture.

To adjust for elevation:

- Move left hand rearward or forward on handguards (moving the hand rearward elevates the muzzle).
- Open or close the "V" of the left hand for small adjustments (closing the "V" elevates the muzzle).

To adjust for a minor cant in the rifle, rotate the handguard left or right in the "V" formed by the thumb and forefinger by rotating the pistol grip left or right.

Crossed Ankle Sitting Position with the Hasty Sling

Figure 5-33. Left View.

Figure 5-34. Front View.

Figure 5-35. Right View.

Crossed Ankle Sitting Position with the Loop Sling

Figure 5-36. Right View.

Figure 5-37. Front View.

Crossed Ankle Sitting Position with the Loop Sling

Apply the three elements and seven factors to this position (para. 5004). To assume crossed ankle sitting position with the loop sling (see figs. 5-36 and 5-37)—

- Position the body at approximately a 30-degree angle to the target.
- Place the left hand under the handguard.
- Bend at knees and break the fall with the right hand.
- Push backward with the feet to extend the legs and place the buttocks on the ground.
- Cross the left ankle over the right ankle.
- Bend forward at the waist and place the left elbow on the left leg below the knee.
- Grasp the rifle butt with the right hand and place the rifle butt into the right shoulder pocket.
- Grasp the pistol grip with the right hand.
- Lower right elbow to the inside of the right knee.
- Lower the head and place the cheek firmly against the stock to allow the aiming eye to look through the rear sight aperture.
- Move the left hand to a location under the handguard, which provides maximum bone support and stability of the weapon.

Crossed Leg Sitting Position with the Hasty Sling

The crossed leg sitting position provides a medium base of support and places some of the body's weight behind the weapon for quick recovery after each shot. Apply the seven factors to this position (para. 5003). To assume the crossed leg sitting position with the hasty sling (see figures 5-38, 5-39, and 5-40)—

- Square the body to the target.
- Grasp the handguard with the left hand.
- Bend at the knees while breaking the fall with the right hand.
- Place the buttocks on the ground as close to the feet as you comfortably can.
- Cross the left leg over the right leg.
- Grasp the pistol grip with the right hand and place the rifle butt in the right shoulder pocket.

Crossed Leg Sitting Position with the Hasty Sling

Figure 5-38. Left View.

Figure 5-39. Front

Figure 5-40. Right View.

- Lower the head and place the cheek firmly against the stock to allow the aiming eye to look through the rear sight aperture.
- Rotate the left hand up, pinching the handguard between the thumb and forefinger.
- Bend forward at the waist while placing the left elbow into the bend of the knee or placing the flat portion of the back of the left arm, just above the elbow, in front of the knee.
- Place the right elbow on the inside of the right knee.
- Adjust the position to adjust sling tension. Squaring the body more to the target or drawing the feet closer together tightens the sling by forcing the left elbow outboard.
- Adjust the position of the left hand to allow the sling to support the weapon and the front sight to be centered in the rear sight aperture.

To adjust for elevation:

- Move left hand rearward or forward on handguards (moving the hand rearward elevates the muzzle).
- Open or close the "V" of the left hand for small adjustments (closing the "V" elevates the muzzle).

To adjust for a minor cant in the rifle, rotate the handguard left or right in the "V" formed by the thumb and forefinger by rotating the pistol grip left or right.

Crossed Leg Sitting Position with the Loop Sling

Apply the three elements and seven factors to this position (para. 5004). To assume crossed leg sitting position with loop sling (see figs. 5-41, 5-42, and 5-43)—

- Position body at a 45- to 60-degree angle to target.
- Place the left hand under the handguard.
- Cross the left leg over the right leg.
- Bend at the knees while breaking the fall with the right hand.
- Place the buttocks on the ground as close to the crossed legs as you comfortably can.
- Bend forward at the waist while placing the left elbow on the left leg into the bend of the knee.
- Grasp the rifle butt with the right hand and place the rifle butt into the right shoulder pocket.
- Grasp the pistol grip with the right hand.
- Lower right elbow to the inside of the right knee.
- Lower the head and place the cheek firmly against the stock to allow the aiming eye to look through the rear sight aperture.
- Move the left hand to a location under the handguard which provides maximum bone support and stability of the weapon.

Open Leg Sitting Position with the Hasty Sling

The open leg sitting position provides a medium base of support and is most commonly used when firing from a forward slope. Apply the seven factors to this position (para. 5003). To assume the open leg sitting position with the hasty sling (see figs. 5-44, 5-45, and 5-46)—

- Square the body to the target.
- Place the feet approximately shoulder width apart.
- Grasp the handguard with the left hand.
- Bend at the knees while breaking the fall with the right hand.
- Push backward with the feet to extend the legs and place the buttocks on the ground.
- Grasp the pistol grip with the right hand and place the rifle butt in the right shoulder pocket.
- Lower the head and place the cheek firmly against the stock to allow the aiming eye to look through the rear sight aperture.

Crossed Leg Sitting Position with the Loop Sling

Figure 5-41. Left View.

Figure 5-42. Front View.

Figure 5-43. Right View.

Open Leg Sitting Position with the Hasty Sling

Figure 5-44. Left View.

Figure 5-45. Front View.

Figure 5-46. Right View.

- Rotate the left hand up, pinching the handguard between the thumb and forefinger.
- Place the flat portion of the back of the left arm, just above the elbow, in front of the left knee.
- Place the right elbow on the inside of the right knee or place the flat portion of the back of the right arm, just above the elbow, in front of the knee.
- Adjust the position to adjust sling tension. Widening the stance tightens the sling by forcing the left elbow outboard. There must be some controlled muscular tension in the legs to hold them up and offer resistance to recoil.
- The position of the left hand to allow the sling to support the weapon and the front sight to be centered in the rear sight aperture.

To adjust for elevation:

- Move left hand rearward or forward on handguards (moving the hand rearward elevates the muzzle).
- Open or close the "V" of the left hand for small adjustments (closing the "V" elevates the muzzle).

To adjust for a minor cant in the rifle, rotate the handguard left or right in the "V" formed by the thumb and forefinger by rotating the pistol grip left or right.

Open Leg Sitting Position with the Loop Sling

Apply the three elements and seven factors to this position (para. 5004). To assume the open leg sitting position with the loop sling (see figs. 5-47, 5-48, and 5-49 on page 5-16)—

- Position the body at approximately a 30-degree angle to the target.
- Place the feet approximately shoulder width apart.
- Place the left hand under the handguard.
- Bend at the knees while breaking the fall with the right hand.
- Push backward with the feet to extend the legs and place the buttocks on the ground.
- Place the left elbow on the inside of the left knee.
- Grasp the rifle butt with the right hand and place the rifle butt into the right shoulder pocket.
- Lower right elbow to the inside of the right knee.
- Lower the head and place the cheek firmly against the stock to allow the aiming eye to look through the rear sight aperture.
- Move the left hand to a location under the handguard which provides maximum bone support and stability of the weapon.

5007. Kneeling Position

Description

The kneeling position is quick to assume and easy to maneuver from. It is usually assumed after initial engagement has been made from a standing position. It can easily be adapted to available cover. A tripod is formed by the left foot, right foot, and right knee when the Marine assumes the position, providing a stable foundation for shooting. The kneeling position also presents a higher profile to facilitate a better field of view as compared to the prone and sitting positions.

Assuming the Kneeling Position

The kneeling position can be assumed by either moving forward or dropping back into position, depending on the combat situation. For example, it may be necessary to drop back into position to avoid crowding cover, or to avoid covering uncleared terrain.

Open Leg Sitting Position with the Loop Sling

Figure 5-47. Left View.

Figure 5-48. Front View.

Figure 5-49. Right View.

Moving Forward into Position

To move forward into the kneeling position, the Marine steps forward toward the target with his left foot and kneels down on his right knee.

Dropping Back into Position

To drop back into the kneeling position, the Marine leaves his left foot in place and steps backward with his right foot and kneels down on his right knee.

High Kneeling Position with the Hasty Sling

Apply the seven factors to this position (para. 5003). To assume the high kneeling position with the hasty sling, either move forward or drop back into position (see figs. 5-50, 5-51, and 5-52):

- Square the body to the target.
- Keep the right ankle straight, with the toe of the boot in contact with the ground and curled under by the weight of the body.
- Place the right portion of the buttocks on or over the right heel. Contact with the heel provides more stability to the position, however, it is not mandatory that the buttocks make contact.
- Place the left foot forward to a point that allows the shin to be vertically straight. For the shin to be vertical, the heel should be directly under the knee. The left foot must be flat on the ground. To provide a wider base of support, slide the right knee and left foot outboard to form a tripod with the right foot.
- Grasp the pistol grip with the right hand and place the rifle butt in the right shoulder pocket.
- Lower the head and place the cheek firmly against the stock to allow the aiming eye to look through the rear sight aperture.
- Rotate the left hand up, pinching the handguard between the thumb and forefinger. The left hand will not grasp the slip ring or the magazine. The magazine must be on the inside of the left arm.
- Place the flat portion of the back of the upper left arm, just above the elbow, on the left knee or against the inside of the left knee so it is in firm

High Kneeling Position with the Hasty Sling

Figure 5-50. Left View.

Figure 5-51. Front View.

Figure 5-52. Right View.

contact. The upper portion of the triceps or the armpit will not rest on the knee.

- Bend the right elbow to provide the least muscular tension possible, lowering it to a natural position.
- Adjust the position to adjust sling tension. Widening the stance by moving the left foot and knee outboard will allow the sling to be tightened.
- Adjust the position of the left hand to allow the sling to support the weapon and the front sight to be centered in the rear sight aperture.

To adjust for elevation:

- Move left hand rearward or forward on handguards (moving the hand rearward elevates the muzzle).
- Open or close the "V" of the left hand for small adjustments (closing the "V" elevates the muzzle).

To adjust for a minor cant in the rifle, rotate the handguard left or right in the "V" formed by the thumb and forefinger by rotating the pistol grip left or right.

High Kneeling Position with the Loop Sling

Apply the three elements and seven factors to this position (para. 5004). To assume the high kneeling position with the loop sling, either move forward or drop back into position (see figs. 5-53, 5-54, and 5-55):

- Position the body at a 45-degree angle to the target.
- Place the left hand under the handguard.
- Kneel down on right knee so right lower leg is approximately parallel to the target (45 to 90 degrees).
- Keep the right ankle straight, with the toe of the boot in contact with the ground and curled under by the weight of the body.
- Place the right portion of the buttocks on or over the right heel; contact with the heel provides more stability to the position.
- Place the left foot forward to a point that allows the shin to be vertically straight. For the shin to be vertical, the heel should be directly under the knee. The left foot must be flat on the ground since it will be supporting the majority of the weight.
- Place the flat portion of the back of upper left arm, just above the elbow, on the left knee so it is in firm contact with the flat surface formed on top of the bent knee. The point of the left elbow will extend just slightly past the left knee. The upper portion of the triceps or the armpit will not rest on the knee.
- Lean slightly forward into the sling for support.
- Grasp the rifle butt with the right hand and place the rifle butt into the right shoulder pocket.
- Grasp the pistol grip with the right hand.
- Bend the right elbow to provide the least muscular tension possible and lower it to a natural position.
- Lower the head and place the cheek firmly against the stock to allow the aiming eye to look through the rear sight aperture.
- Move the left hand to a location under the handguard, which provides maximum bone support and stability for the weapon.

Medium Kneeling Position

This is also referred to as the bootless kneeling position. Assume the medium kneeling position in the same way as the high kneeling position with the ex-

High Kneeling Position with the Loop Sling

Figure 5-53. Left View.

Figure 5-54. Front View.

Figure 5-55. Right View.

ception of the right foot. The right ankle is straight and the foot is stretched out with the bootlaces in contact with the ground. The buttocks are in contact with the heel of the right foot. See figure 5-56.

Figure 5-56. Medium Kneeling Position.

Low Kneeling Position

The low kneeling position is most commonly used when firing from a forward slope. Assume the low kneeling position in the same way as the high kneeling position with the exception of the placement of the right foot. Turn the right ankle so the outside of the foot is in contact with the ground and the buttocks are in contact with the inside of the foot. See figure 5-57.

Figure 5-57. Low Kneeling Position.

5008. Standing Position

Description

The standing position is the quickest position to assume and the easiest to maneuver from. It allows greater mobility than other positions. The standing position is often used for immediate combat engagement. The standing position is supported by the shooter's legs and feet and provides a small area of contact with the ground. In addition, the body's center of gravity is high above the ground. Therefore, maintaining balance is critical in this position.

Standing Position with the Hasty Sling

Apply the seven factors to this position (para. 5003). To assume the standing position with the hasty sling (see figs. 5-58, 5-59, and 5-60)—

- Square the body to the target.
- Spread the feet apart to a comfortable distance with the left foot slightly in front of the right foot. This distance may be wider than shoulder width.
- Distribute the weight evenly over both feet and hips. Balance will shift forward slightly to reduce recovery time and increase the stability of the hold. The legs should be slightly bent for balance.
- Grasp the pistol grip with the right hand and place the rifle butt in the right shoulder pocket.
- Bring the rifle sights up to eye level instead of lowering the head to the sights and place the cheek firmly against the stock. Ensure the head is erect so the aiming eye can look through rear sight aperture.
- Rotate the left hand up, pinching the handguard between the thumb and forefinger.
- The left hand will be under the handguard with the thumb on the outboard side of the handguard.The left hand will not grasp the slip ring or magazine. The magazine must be on the inside of the left arm.
- Hold the right elbow in a natural position.
- Adjust the position of the left hand on the handguard to allow the sling to support the weapon and front sight to be centered in the rear sight aperture.

To adjust for elevation:

- Move left hand rearward or forward on handguards (moving the hand rearward elevates the muzzle).
- Open or close the "V" of the left hand for small adjustments (closing the "V" elevates the muzzle).
- Adjust the position to adjust sling tension. Moving the left elbow out or squaring the body more to the target tightens the sling.

To adjust for a minor cant in the rifle, rotate the handguard left or right in the "V" formed by the thumb and forefinger by rotating the pistol grip left or right.

Standing Position with the Hasty Sling

Figure 5-58. Left View.

Figure 5-59. Front View.

Figure 5-60. Right View.

Standing Position with the Parade Sling

The parade sling is used to emphasize marksmanship fundamentals while firing from the standing position on KD courses during entry-level training. To achieve proper positioning of the parade sling, perform the following steps:

- Attach sling to rifle by placing feed end of sling down through the upper sling swivel.
- Place feed end of sling through sling keeper and lock into place.
- Attach J-hook to lower sling swivel.
- Pull feed end of sling through sling keeper until sling is taut.
- Move sling keeper down near feed end of sling.
- Lock sling keeper into place.

Perform the following steps to assume the standing position with a parade sling (figure 5-61):

- Stand erect.
- Face approximately 90 degrees to the right of the line of fire.
- Place feet approximately shoulder width apart.
- Place left hand under handguard.
- Grasp pistol grip with right hand.
- Place rifle butt into right shoulder pocket.
- Invert left elbow across rib cage.
- Rest left arm naturally against rib cage.
- Lower right elbow to a natural position.
- Place cheek firmly against stock to obtain a firm stock weld.

Figure 5-61. Standing Position with the Parade Sling.

CHAPTER 6. USE OF COVER AND CONCEALMENT

Note

+The procedures in this manual are written for right-handed Marines; left-handed Marines should reverse instructions as necessary.

6001. Cover and Concealment

In a combat environment, a Marine must be prepared to fire from any type of cover or concealment. Cover is anything that protects a Marine from enemy fire. Cover may be an existing hole, a hastily dug shelter, or a well-prepared fighting position with overhead protection. Concealment is anything that hides a Marine from enemy view, but it may not afford protection. Concealment can be obtained from buildings, trees, crops, and skillful use of ground contours. A Marine can use any object or terrain feature that protects him from enemy fire, hides him from enemy view, allows him to observe the enemy, and provides support for a firing position.

Types of Cover

Frontal Cover

A firing position should have frontal cover that provides protection from small arms fire and indirect fire fragments. Ideally, frontal cover should be thick enough to stop small arms fire, high enough to provide protection from enemy fire, and wide enough to provide cover when firing to the left or right edge of a sector of fire.

Ideal Cover

The ideal cover provides:

- Overhead, flank, and rear protection from direct and indirect fire.
- Free use of personal weapons.
- Concealment from enemy observation.
- A concealed route in and out.
- Unobstructed view of the sector of fire.

Common Cover Materials

Any material that protects a Marine from small arms fire can be used for cover. Some common materials include sandbags, trees, logs, and building debris. Table 6-1 presents some common materials and their minimum thickness required for protection from small arms fire.

Table 6-1. Minimum Thickness for Protection Against Small Arms.

Material	Minimum Thickness (in inches)
Concrete	7
Broken stone (rubble)	20
Dry sand	24
Wet sand	35
Logs (oak)	40
Earth (packed)	48

Sandbags

Cover can be improved and positions can be fortified by filling sandbags with dirt/sand and placing them around the position. Sandbags should be tightly packed because bullets can easily penetrate moist or loosely packed sandbags. Overlapping sandbags increase protection and decrease the bullet's ability to penetrate the sandbag. A minimum thickness of three sandbags is required to stop small arms fire.

Trees/Logs

Wood is a relatively dense material and offers good cover and protection. Bullets have a tendency to fragment when they penetrate wood. Live trees have a greater resistance to bullet penetration than dead trees. Wood that has been treated with creosote, such as telephone poles and railroad ties, offers better protection from projectiles than untreated wood, but it still does not ensure protection from small arms fire.

Cinder Bocks

Cinderblocks are not impenetrable cover. Although they are made of a dense material, the composition of a cinderblock is so brittle that a bullet can shatter the block upon impact. This can cause injury to a Marine by secondary fragmentation.

Firing From Specific Types of Cover

Effective cover allows a Marine to engage enemy targets while protecting himself from enemy fire. Several types of cover provide support, protection, and concealment and do not interfere with target engagement. A Marine must adapt firing positions to the type of cover available.

Fighting Hole

A Marine should use fighting holes if available. See figure 6-1. After a Marine enters the fighting hole, he adds or removes dirt, sandbags or other supports to fit his height. To assume a firing position, a Marine performs the following steps:

- Place the right foot to the rear as a brace.
- Lean forward until the chest is against the forward wall of the fighting hole.
- Extend the left arm and elbow over the forward side of the fighting hole so the left forearm rests against the back of the parapet.
- Place the rifle butt into the pocket of the right shoulder and grasp the pistol grip with the right hand.
- Place the right elbow on solid support using the elbow rest of the fighting hole or sandbags placed around the fighting hole.

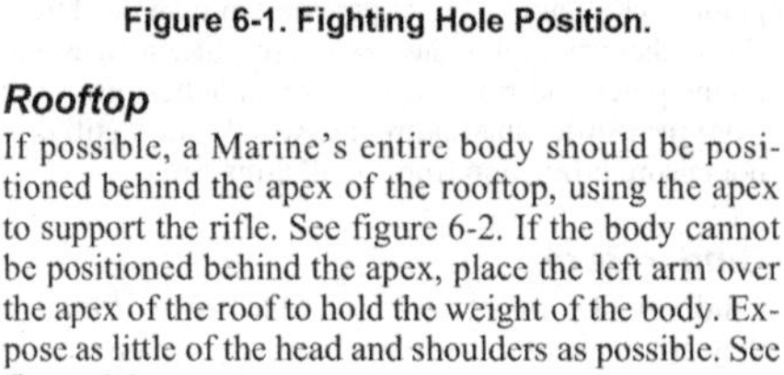

Figure 6-1. Fighting Hole Position.

Rooftop

If possible, a Marine's entire body should be positioned behind the apex of the rooftop, using the apex to support the rifle. See figure 6-2. If the body cannot be positioned behind the apex, place the left arm over the apex of the roof to hold the weight of the body. Expose as little of the head and shoulders as possible. See figure 6-3.

Figure 6-2. Rooftop Position Behind the Apex.

Figure 6-3. Rooftop Position Supported by the Apex.

Window

The Marine can establish a supported or unsupported position from a window.

Unsupported. A Marine can establish an unsupported position back from the opening of the window so that the muzzle does not protrude and interior shadows provide concealment so as not to provide a silhouette to the enemy. See figure 6-4.

Figure 6-4. Unsupported Window Position.

Supported. When additional stability is needed, a Marine can establish a supported position by placing the rifle handguards or his forearm in the "V" formed by the side and bottom of the windowsill. A drawback to this technique is that the muzzle of the weapon and the Marine may be exposed to view. See figures 6-5 and 6-6.

Vehicle

In many combat situations, particularly in urban environments, a vehicle may be the best form of cover. When using a vehicle for cover, the engine block provides the most protection from small arms fire. The Marine should establish a position behind the front wheel so the engine block is between him and the target (see figs. 6-7 and 6-8). From this position, the Marine may fire over, under or around the vehicle. This is a particularly effective position for larger vehicles that are high off the ground.

The Marine can establish additional support for the rifle by positioning himself behind the doorjamb (frame of door) and placing the rifle against the "V" formed by the open door and doorframe (see figure 6-9). From this position, the Marine may fire over the hood of the vehicle while using the engine block for protection.

Figure 6-5. Supported Window Position (Handguards on Support).

Figure 6-6. Supported Window Position (Forearm on Support).

Figure 6-7. Firing Around Front of Vehicle.

Figure 6-8. Firing Over Front of Vehicle.

Figure 6-9. Establishing a Supported Position in a Vehicle.

However, this position limits lateral mobility and it is more difficult to maneuver from.

At the back of the vehicle, only the axle and the wheel provide cover. If the Marine must shoot from the back of the vehicle, he must position himself directly behind the wheel as much as possible (see fig. 6-10).

Figure 6-10. Using the Back of a Vehicle for Cover.

6002. Supported Firing Positions

Supports are foundations for positions; positions are foundations for the rifle. To maximize the support the position provides, the firing position should be adjusted to fit or conform to the shape of the cover. Elements of a sound firing position, such as balance and stability, must be incorporated and adjusted to fit the situation and type of cover. A supported firing position should minimize exposure to the enemy, maximize the stability of the rifle, and provide protection from enemy observation and fires. A Marine can use any available support (e.g., logs, rocks, sandbags or walls) to stabilize his firing position. The surrounding combat environment dictates the type of support and position used.

Considerations Using Cover and Concealment

Cover and concealment considerations are similar, regardless of the combat environment (e.g., urban, desert, jungle).

Adjusting the Shooting Position

The type of cover can dictate which shooting position (e.g., standing, kneeling, sitting, and prone) will be the most effective. For example, a Marine's height in relation to the height of the cover aids in the selection of a firing position.

The firing position selected should be adjusted to fit the type of cover to:

- Provide stability. The position should be adjusted to stabilize the rifle sights and allow the management of recoil to recover on target.
- Permit mobility. The position should be adjusted to permit lateral engagement of dispersed targets and movement to other cover.
- Allow observation of the area/enemy while minimizing exposure to the enemy.

The firing position is adjusted to fit the type of cover by adjusting the seven factors (i.e., left hand, pocket of shoulder, right elbow, stock weld, grip of right hand) to support the rifle or the position.

Keeping the Entire Body Behind Cover

A Marine should minimize exposure of any part of his body to fire. Be especially aware of the head, right elbow, knees or any other body part that may extend beyond the cover.

Firing From the Right or Left Side of Cover

To minimize exposure and maximize the cover's protection, a right-handed Marine should fire from the right side of cover and a left-handed Marine should fire from the left side, if possible (see fig. 6-11).

Figure 6-11. Firing from the Right Side of Cover.

Figure 6-12. Firing from the Left Side of Cover.

If, however, a right-handed Marine must fire from the left side of cover, he fires right-handed but adjusts his position behind cover and uses the rollout technique (see para. 6003) to engage the target. See figure 6-12.

Firing Over the Top of Cover

Firing over the top of cover provides a wider field of view and lateral movement. When firing over the top of cover, the position may be supported and stabilized by resting the handguard or the left forearm on the cover (see fig. 6-13). The Marine should keep as low a profile as possible; the rifle should be as close to the top of cover as possible.

Figure 6-13. Firing Over the Top of Cover.

Maintaining Muzzle Awareness

When firing over the top of cover, a Marine must remember that the sights are higher than the barrel and remain aware of the location of his muzzle. Therefore, a Marine must maintain a position that ensures the muzzle is high enough to clear the cover (e.g., window sill, top of wall) as he obtains sight alignment/sight picture on the target (see figs. 6-14a and 6-14b).

Clearing the Ejection Port

Ensure the cover does not obstruct the ejection port. If the ejection port is blocked, the obstruction can interfere with the ejection of the spent cartridge case and cause a stoppage.

Figures 6-14a and 6-14b. Clearing Cover with the Muzzle.

Resting the Magazine

The bottom, front or side of the rifle magazine can rest on or against support to provide additional stability (see figs. 6-15, 6-16, and 6-17).

CAUTION

The back of the magazine should not be pulled back against support because it can cause a stoppage by not allowing a round to feed from the magazine.

Seven Factors

The seven factors (left hand, rifle in shoulder pocket, stock weld, right elbow, grip of the right hand, breathing, muscular tension) are applied when firing from cover, however, some may have to be modified slightly to accommodate the artificial support provided the rifle and position.

Left Hand

The support should be used to help stabilize both the firing position and the rifle to enable the Marine to maintain sight alignment and sight picture.

The forearm or left hand can contact the support to stabilize the weapon. Rest the forearm or the meaty portion of the bottom of the left hand on the support and rest the rifle in the "V" formed by the thumb and forefinger of the left hand (see fig. 6-18).

The rifle's handguards may rest on the support, but the barrel may not (see fig. 6-19). Placement of left hand on the handguard may have to be adjusted forward or back to accommodate the cover and the additional support provided by the rifle resting on the cover. If the handguards are resting on the cover, the left hand

Figure 6-15. Bottom of Magazine on Support.

Figure 6-16. Front of Magazine on Support.

Figure 6-17. Side of Magazine on Support.

can pull down on the handguards to further stabilize the weapon.

Rifle in Shoulder Pocket

Regardless of the combat situation or the height of the cover, the rifle butt must remain in the shoulder pocket to manage the effects of recoil, stabilize the rifle, and maintain the rifle's battlesight zero. The firing position must be adjusted behind cover to enable the rifle to be placed in the shoulder.

Stock Weld

Regardless of the cover, the firing position must be adjusted to allow stock weld to be achieved. Proper stock weld provides quick recovery between shots and keeps the aiming eye centered in the rear sight aperture.

Right Elbow

The right elbow can be placed on or against the support to stabilize the weapon and the position. Ensure the elbow is not extended beyond cover revealing the position to the enemy.

Grip of the Right Hand

If rifle handguards, forward hand, or forearm rest on cover for support, the grip of the right hand should pull back and down on the pistol grip to further stabilize the weapon in the shoulder and on the support.

Breathing

Breathing does not change when firing from a supported position.

Muscular Tension

To create balance and support for the position, the Marine may shift his body weight into or against the support (see fig. 6-20). This enables the Marine to use cover to support his body weight, reducing the need for muscular tension.

Figure 6-18. Forearm Resting on Cover.

Figure 6-19. Handguards Resting on Cover.

Figure 6-20. Shifting Body Weight into Cover.

Types of Supported Positions

Supported Prone

If possible, a Marine should use the supported prone position when firing from behind cover. This position is the steadiest, provides the lowest silhouette, and provides maximum protection from enemy fire.

- Support the position by placing the handguards, the forearm, or the magazine on or against support (see fig. 6-21).

Figure 6-21. Supported Prone.

- The prone position can be assumed behind a tree, a wall, a log, or almost any type of cover. It is flexible and allows shooting from all sides of cover and from cover of various sizes.
- The body must be adjusted to conform to the cover. For example, if the cover is narrow, keep the legs together. The body should be in line with the rifle and directly behind the rifle (see figs. 6-22 and 6-23). This presents a smaller target to the enemy and provides more body mass to absorb recoil.

Figure 6-22. Supported Prone Behind Narrow Cover—Rear View.

Supported Kneeling

When the prone position cannot be used because of the height of the support, the supported kneeling position may be appropriate. The kneeling position provides additional mobility over the prone position.

- The kneeling position allows shooting from all sides and from cover of varying sizes. This position may be altered to maximize the use of cover or support by assuming a variation of the kneeling position (high, medium or low).
- In the kneeling position, the Marine must not telegraph his position behind the cover with his knee. When shooting around the sides of cover, the Marine should strive to keep his right knee in line with his left foot so as not to reveal the position to the enemy. See figure 6-24.
- Support the position by placing the handguards, the forearm, or the magazine on or against support. In addition, the position (e.g., a knee, the side of the body) may rest against support (see fig. 6-25).
- If the rifle is resting on support, the Marine may not need to stabilize the weapon by placing his left elbow on his knee (see fig. 6-26).

Supported Sitting

A supported sitting position may be used to fire over the top of cover when mobility is not as critical. A sitting position can be comfortably assumed for a longer period of time than a kneeling position and it can conform to higher cover when a prone position cannot be used.

Figure 6-23. Supported Prone Behind Narrow Cover—Front View.

Figure 6-24. Supported Kneeling.

Figure 6-25. Supported Kneeling with Body Against Support.

- Support the position by placing the handguards, the forearm, or the magazine on or against support (see fig. 6-27).

- If the rifle is resting on support, the Marine may not need to stabilize the weapon by placing his left or right elbows on his legs (see fig. 6-28).

Supported Standing

The supported standing position provides greater mobility than the other positions and usually provides greater observation of the enemy. The supported standing position is effectively used behind high cover (e.g., window, over a wall) or narrow cover (e.g., tree, telephone pole).

- To assume the supported standing position, the Marine leans his body forward or against support to stabilize the weapon and the position.

Figure 6-26. Supported Kneeling with Rifle Resting on Support.

Figure 6-27. Supported Sitting.

Figure 6-28. Supported Sitting with Rifle on Support.

- Support the position by placing the handguards, the forearm, or the magazine on or against support. In addition, the position (e.g., the side of the body) may rest against support (see fig. 6-29).

Figure 6-29. Supported Standing.

6003. Searching for and Engaging Targets From Behind Cover

To locate targets when behind cover or to ensure the area is clear before moving, the Marine must expose as little of himself as possible to the enemy. Additionally, the Marine must be ready to fire if a target is located. Two techniques that can be used to locate and engage targets from behind cover are: the pie and rollout. These techniques minimize the Marine's exposure to enemy fire while placing the Marine in a position to engage targets or to move to another location if necessary. These techniques are also used to enter a building or structure. Both techniques are used in the kneeling and standing positions. To be accurate in engaging targets using either technique, the seven factors must be applied. See page 6-6 for a discussion of seven factors.

Pie Technique

To perform the pie technique:

- Staying behind cover, move back and away from the leading edge of the cover. The surroundings and situation will dictate the distance you should move back and away from the cover. Generally, the further back the Marine is from cover, the greater his area of observation; staying too close to cover decreases the area of observation. However, if the Marine is too far back from the leading edge of cover, he may unknowingly expose himself to the enemy.
- Assume a firing position and lower the rifle sights enough to have a clear field of view, orienting the muzzle on the leading edge of the cover. (In a building, the baseboards serve as a reference point for the muzzle of the rifle when searching for targets.)
- Taking small side steps, slowly move out from behind the cover, covering the field of view with the aiming eye and muzzle of the weapon. Wherever the eyes move, the muzzle should move (eyes, muzzle, target). The muzzle should remain on the leading edge of cover, serving as a pivot point when moving out.
- Continue taking small side steps and moving out from cover until a target is identified or the area is found to be clear.
- When a target is identified, sweep the safety, place the finger on the trigger, and engage the target.

Rollout Technique

To perform the rollout technique:

- Staying behind cover, move back and position the body so it is in line with the leading edge of the cover, ensuring that no part of the body extends beyond the cover.
- Assume a firing position and come to the Ready, ensuring the muzzle is just behind the cover.
- Canting the head and weapon slightly, roll the upper body out to the side just enough to have a clear field of view and allow the muzzle to clear the cover. Keeping the feet in place, push up on the ball of one foot to facilitate rolling out.
- Continue rolling out from cover until a target is identified or the area is found to be clear.
- When a target is identified, sweep the safety, place the finger on the trigger, and engage the target.
- If a target is identified before moving out from cover, the rifle should be taken off safe before moving out using the rollout technique.

Combining Techniques

In some situations, it may be necessary to combine the pie and rollout techniques to search an entire area for targets (e.g., corner of a building, a doorway). Changing from one technique to another may permit the Marine to minimize his exposure to enemy observation and fires.

6004. Moving Out From Behind Cover

A Marine must be constantly aware of his surroundings and available cover, should a threat appear. He should avoid obvious danger areas and move quickly through danger areas that cannot be avoided.

When moving from cover to cover, the Marine should select the next covered location and plan his route before moving from his present position. This is done by quickly looking from behind cover to ensure the area is clear, ensuring the head and eyes are exposed for as short a time as possible.

If necessary, the Marine should conduct a Condition 1 reload before moving from cover.

Once the Marine is committed to moving, all focus should be on moving until cover is reassumed.

CHAPTER 7. RIFLE PRESENTATION

Note

☞The procedures in this manual are written for right-handed Marines; left-handed Marines should reverse instructions as necessary.

7001. Presentation of the Rifle

In a combat environment, targets may present themselves with little or no warning. To maintain an advantage, the Marine carries his weapon in a position appropriate to the threat level that permits the rifle to be both easily carried and presented as quickly as possible. A carry is also established based on the situation such as moving in a close quarter environment, moving over or under objects, etc.

Presenting the Rifle From the Tactical Carry

The Marine uses the Tactical Carry when no immediate threat is present. This carry permits the rifle to be easily carried for long periods of time, but it does not permit the quickest presentation to a target. If the situation changes and a target presents itself, a Marine performs the following steps to present the rifle from the Tactical Carry once a target appears:

- Extend the rifle toward the target keeping the muzzle slightly up so the buttstock clears all personal equipment. Continue to look at the target.
- At the same time, place the rifle in Condition 1. Two methods can be used to place rifle in Condition 1 if it is in Condition 3:
  - Grip the pistol grip firmly with the right hand. Pull the charging handle with the left hand to its rearmost position and release (see fig. 7-1).

Figure 7-1. Pulling Charging Handle with Left Hand.

  - Grip the handguards firmly with the left hand. Pull the charging handle with the right hand to its rearmost position and release (see fig. 7-2).

Figure 7-2. Pulling Charging Handle with Right Hand.

- As the rifle is being presented, take the rifle off safe and place the trigger finger on the trigger (see figs. 7-3a, 7-3b, and 7-3c on page 7-2).
- Level the rifle while pulling it firmly into the pocket of the shoulder to obtain proper stock weld. Do not move the head down to meet the stock of the rifle.

Note

If the rifle is in the shoulder properly, the aiming eye will be able to look through the rear sight as soon as the stock makes contact with the cheek.

- As the sights become level with the aiming eye, visually locate the target through the rear sight aperture. As the rifle sights settle, shift the focus back to the front sight post to obtain sight alignment, and place the tip of the post center mass on the target to obtain sight picture.

Presenting the Rifle From the Alert Carry and From the Ready Carry

The Marine uses the Alert Carry when enemy contact is likely. The Alert is also used for moving in close terrain (e.g., urban, jungle).

The Marine uses the Ready Carry when enemy contact is imminent.

Figures 7-3a, 7-3b, and 7-3c. Sweeping the Safety.

To present the rifle from the Alert and from the Ready, a Marine performs the following steps once a target appears:

- While looking at the target, bring the muzzle up by raising the left hand, allowing the rifle butt to pivot in the shoulder. At the same time, pull the rifle firmly into the pocket of the shoulder.
- As the rifle is being presented, take the rifle off safe and place the trigger finger on the trigger (see figs. 7-3a, 7-3b, and 7-3c).
- As the stock makes contact with the cheek, level the rifle to obtain proper stock weld.
- Do not move the head down to meet rifle stock.

Note

If the rifle is in the shoulder properly, the aiming eye will be able to look through the rear sight as soon as the stock makes contact with the cheek.

- As the sights become level with the aiming eye, visually locate the target through the rear sight aperture. As the rifle sights settle, shift the focus back to the front sight post to obtain sight alignment, and place the tip of the post center mass on the target to obtain sight picture.

Presenting the Rifle From the Strong Side Sling Arms Transport

Once a target appears, a Marine performs the following to present the rifle from Strong Side Sling Arms:

- While looking at the target, lean forward slightly to facilitate removal of the rifle from the shoulder.
- Reach under the right arm with the left hand between the sling and the body and grasp the handguards. See figure 7-4). At the same time, pull down on the sling and raise right elbow out and parallel to the deck.

Figure 7-4. Grasping the Handguards.

- Roll the right shoulder forward and release the sling from the right hand once the handguards have cleared the elbow. At the same time, pull rifle forward off the shoulder with left hand.
- Continue pulling the rifle forward with the left hand while rotating the rifle parallel to the deck; when the right arm is free of the sling and the rifle clears all personal gear, grasp the charging handle with the right hand (see fig. 7-5) and pull it to its rearmost position and release.
- Establish a firing grip with right hand while keeping the trigger finger straight along the receiver.
- Take rifle off safe and place the trigger finger on the trigger (see figs. 7-3a, 7-3b, and 7-3c).
- Level the rifle while pulling it firmly into the pocket of the shoulder to obtain proper stock weld. Do not move the head down to meet the stock of the rifle.

Figure 7-5. Clearing Gear and Grasping the Charging Handle.

Note

If the rifle is in the shoulder properly, the aiming eye will be able to look through the rear sight as soon as the stock makes contact with the cheek.

- As the sights become level with the aiming eye, visually locate the target through the rear sight aperture. As the rifle sights settle, shift the focus back to the front sight post to obtain sight alignment, and place the tip of the post center mass on the target to obtain sight picture.

Presenting the Rifle From the Weak Side Sling Arms Transport

The hasty sling should be maintained while presenting the rifle from this transport. To present the rifle from Weak Side Sling Arms, a Marine performs the following steps once a target appears:

- While looking at the target, lean forward slightly to facilitate removal of rifle from the shoulder.
- Grasp the sling with the right hand to prevent the rifle from falling off the shoulder.
- Grasp handguards with left hand (the index finger points toward the muzzle). See figure 7-6.
- Rotate the rifle counterclockwise while extending the muzzle toward the target.
- Continue extending the rifle toward the target to ensure the rifle clears all personal gear.
- Grasp the charging handle with the right hand and pull it to its rearmost position and release (see fig. 7-7).
- Establish a firing grip with right hand while keeping the trigger finger straight along the receiver.
- Take the rifle off safe and place the trigger finger on the trigger (see figures 7-3a, 7-3b, and 7-3c).
- Level the rifle while pulling it firmly into the pocket of the shoulder to obtain proper stock weld. Do not move the head down to meet the stock of the rifle.

Figure 7-6. Grasping the Handguards.

Note

If the rifle is in the shoulder properly, the aiming eye will be able to look through the rear sight as soon as the stock makes contact with the cheek.

Figure 7-7. Clearing Gear and Grasping the Charging Handle.

- As the sights become level with the aiming eye, visually locate the target through the rear sight aperture. As the rifle sights settle, shift the focus back to the front sight post to obtain sight alignment, and place the tip of the post center mass on the target to obtain sight picture.

7002. Search and Assess

After a Marine engages a target, he must immediately search the area and assess the results of his engagement. Searching and assessing enables the Marine to avoid tunnel vision that can restrict the focus so that an indication of other targets is overlooked.

Purpose

The Marine searches the area for additional targets or for cover. The Marine assesses the situation to determine if he needs to re-engage a target, engage a new target, take cover, assume a more stable position, cease engagement, etc.

Technique

To search and assess, a Marine performs the following steps:

- Keeping the buttstock in the shoulder, lower the muzzle of the rifle slightly to look over the sights.
- Place the trigger finger straight along the receiver (see fig. 7-8).
- Search the area and assess the situation/threat by moving the head, eyes, and rifle left and right (approximately 45 degrees from center) to cover the immediate area. The muzzle moves with the head and eyes in one fluid motion while searching. Keep both eyes open to increase the field of view.

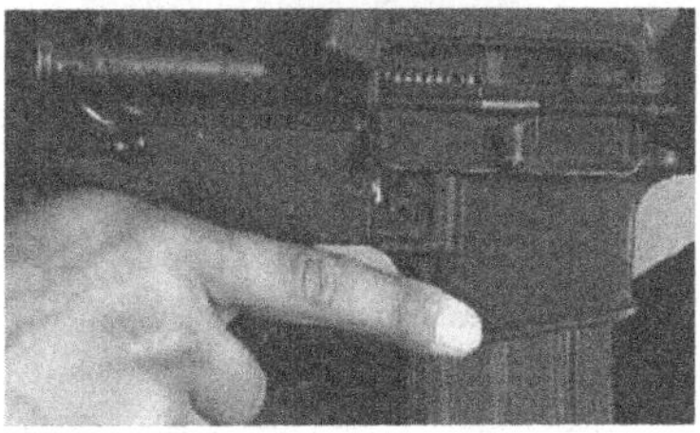

Figure 7-8. Straight Trigger Finger.

- Once a Marine determines the area is clear of enemy threat, he places the rifle on safe.

Searching and Assessing to a Higher Profile

Depending on the tactical situation, the Marine may choose to increase his area of observation by searching and assessing to a higher profile position.

Prone to Kneeling

After searching and assessing at the prone position, move to a kneeling position by performing the following steps:

- While maintaining control of the pistol grip, lower the rifle butt out of the shoulder.
- Drop the left hand to the deck and, bringing it back, push up off the deck to both knees (see figs. 7-9a and 7-9b).
- Grasp the handguard with the left hand and place the rifle butt in the pocket of the shoulder.
- Assume a kneeling position and search and assess (see fig. 7-10).

Sitting to Kneeling

After searching and assessing at the sitting position, move to a kneeling position by performing the following steps:

- Maintain control of the rifle with the rifle butt in the pocket of the shoulder.

Figures 7-9a and 7-9b. Pushing Up Off the Deck to Both Knees.

- Uncross the legs to an open leg position.
- Tuck the right foot underneath the left thigh, as close to the buttocks as possible (see fig. 7-11).
- Lean forward and to the right and roll on to the right knee to a kneeling position and search and assess (see fig. 7-12).

Note

It may be necessary to release the rifle with the right hand and push off the deck with the right hand to assist in rolling up to a kneeling position.

Kneeling to Standing

After searching and assessing at the kneeling position, maintain control of the rifle with the rifle butt in the pocket of the shoulder, and stand while continuing to search and assess.

Figure 7-10. Kneeling Search and Assess.

Figure 7-11. Tucking the Right Foot.

Figure 7-12. Rolling Up to a Kneeling Position.

CHAPTER 8. EFFECTS OF WEATHER

Wind, temperature, and precipitation can affect the trajectory of the bullet. In addition, all weather conditions have a physical and psychological effect on Marines. Marines must use techniques to offset the effects of wind, light, temperature, and precipitation (snow, sleet, and rain). Through proper training, Marines can develop the confidence required to reduce the physical and psychological effects of weather during combat situations.

8001. Physical Effects of Wind on the Bullet

Physical Effects

The weather condition that presents the greatest problem to shooting is the wind. Wind affects a bullet's trajectory. The effect of wind on the bullet as it travels down range is referred to as deflection. The wind deflects the bullet laterally in its flight to the target (see fig. 8-1).

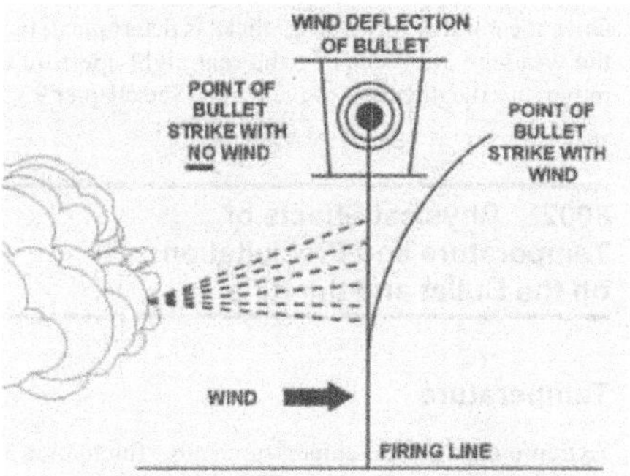

Figure 8-1. Deflection of a Bullet.

The bullet's exposure time to the wind determines the amount the bullet is deflected from its original trajectory. Deflection increases as the distance to the target increases. There are three factors that affect the amount of deflection of the bullet:

- Velocity of the wind—The greater the velocity of the wind, the more the bullet will be deflected.
- Range to the target—As the distance to the target increases, the speed of the bullet slows allowing the wind to have a greater effect on shot placement.
- Velocity of the bullet—A bullet with a high muzzle velocity will not be affected by the wind as much as a bullet with a low muzzle velocity.

Determining Windage Adjustments to Offset Wind Effects

The velocity and direction of the wind in relationship to the bullet must be determined to offset the wind's effects. If Marines can classify wind values and determine velocity within 5 mph, they can effectively engage targets in windy conditions.

Wind Direction

Determine wind direction by observing direction vegetation is moving, by feeling the wind blow against the body, or by observing direction of a flag (in training).

Wind Value Classifications

Winds are classified according to the direction from which they are blowing in relation to the direction of fire. The clock system indicates wind direction and value (see fig. 8-2). Winds can be classified as half

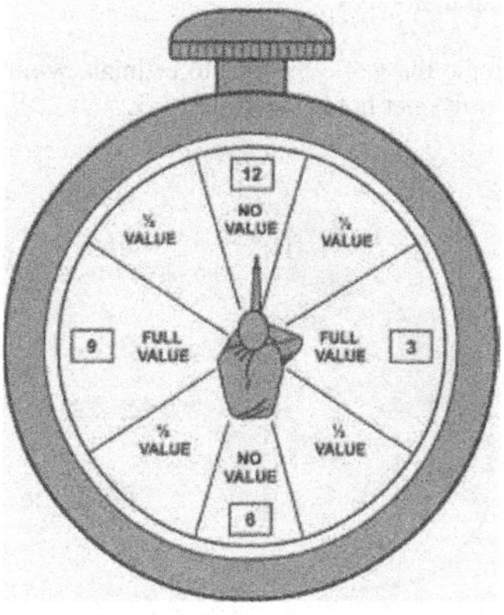

Figure 8-2. Clock System.

value, full value, or no value. The target is always located at 12 o'clock.

Wind Velocity

There are two methods used to determine wind velocity: observation and flag. The flag method is used as a training tool on the known distance (KD) range to learn the observation method. This method teaches Marines to relate the effect a given wind condition has on the natural surroundings in order to develop the base of knowledge used during the observation method. The observation method is the primary method used to estimate wind velocity and direction in a tactical situation. The following are guidelines used during the observation method:

- Under 3 miles per hour (mph) the wind can hardly be felt on the face. The presence of a slight wind can be detected by drifting smoke.
- 3 to 5 mph winds can be felt lightly on the face.
- 5 to 8 mph winds keep tree leaves in a constant motion.
- 8 to 12 mph winds raise dust and loose paper.
- 12 to 15 mph winds cause small trees to sway.
- 16 to 25 mph winds cause large trees to sway.

Flag Method

The flag method is primary method used on the KD range. To estimate wind velocity in miles per hour:

Estimate the angle created between the flagpole and the flag in degrees.

- Divide the angle by four to estimate wind velocity in miles per hour. See figure 8-3.

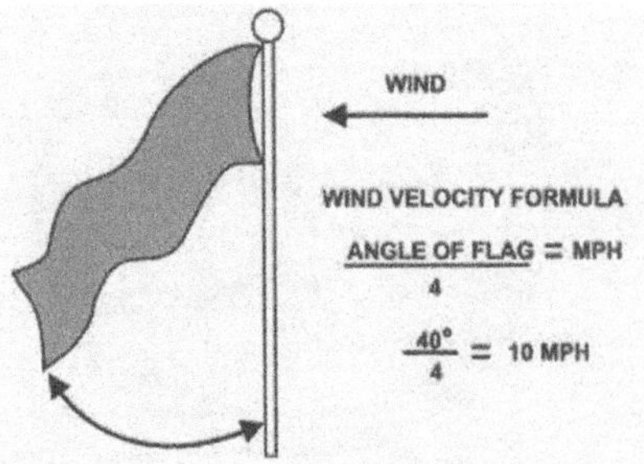

Figure 8-3. Flag Method.

Note

Information given is based on a dry flag. A wet flag is heavy and gives a false reading.

Windage Adjustments

After identifying wind direction, wind classification, and wind velocity, windage adjustments needed to enable the bullet to strike the target are estimated in the following ways:

Observation Method. Using the windage chart provided in figure 8-4, match the wind velocity, wind direction, and range to the target to the information in the chart to estimate the correct number of clicks to apply to the windage knob.

Flag Method. Using the windage chart provided in figure 8-5, match the wind velocity, wind direction, and range to the target to the information in the chart to determine the correct number of clicks to apply to the windage knob.

Once the number of windage clicks is determined, turn the windage knob causing the rear sight aperture to move into the direction of the wind. (See chapter 9.)

8002. Physical Effects of Temperature and Precipitation on the Bullet and the Rifle

Temperature

Extreme changes in temperature cause fluctuation in the rifle's chamber pressure. This fluctuation is caused by changes in the propellant's temperature. In cold weather, as rifle chamber pressure decreases, the bullet exits the muzzle at a lower velocity, and the bullet impacts the target below the point of aim. In extreme heat, the rifle's chamber pressure increases causing the bullet to exit the muzzle at a higher velocity and impact the target above the point of aim. Hot air is less dense than cool air and provides less resistance to the bullet; this allows the bullet to travel faster and experience less deflection from the wind. Cold air is dense and provides the bullet with more resistance; this causes the bullet to travel slower and experience greater deflection from the wind.

	WINDS CAN BE FELT LIGHTLY ON FACE AND LEAVES ARE IN CONSTANT MOTION		WINDS RAISE DUST AND LOOSE PAPER 8-12 MPH		WINDS CAUSE SMALL TREE TO SWAY 12-15 MPH		WINDS CAUSE LARGE TREES TO SWAY 16-20 MPH		21-25 MPH	
	WIND VALUE		WIND VALUE		WIND VALUE		WIND VALUE		WIND VALUE	
RANGE YARDS	FULL	HALF	FULL	HALF	FULL	HALF	FULL	HALF	FULL	HALF
200	2	1	3	1	5	2	6	3	8	4
300	3	1	6	3	10	5	13	6	16	8
400	4	2	9	4	13	6	18	9	22	11
500	6	3	12	6	18	9	24	12	30	15

Figure 8-4. Windage Click Chart for the Observation Method.

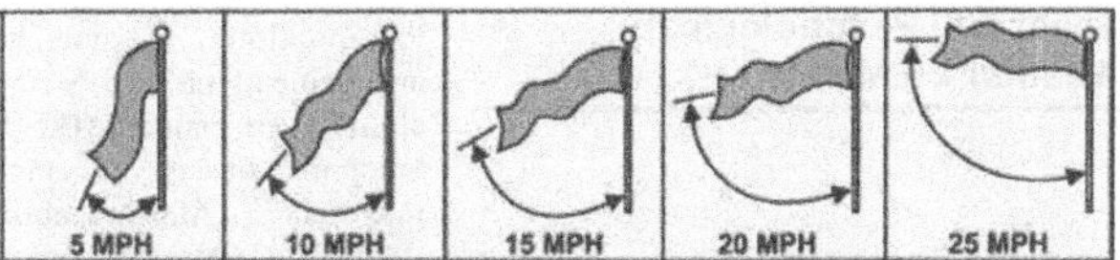

RANGE FLAG ANGLES	5 MPH		10 MPH		15 MPH		20 MPH		25 MPH	
	WIND VALUE		WIND VALUE		WIND VALUE		WIND VALUE		WIND VALUE	
RANGE YARDS	FULL	HALF	FULL	HALF	FULL	HALF	FULL	HALF	FULL	HALF
200	2	1	3	1	5	2	6	3	8	4
300	3	1	6	3	10	5	13	6	16	8
400	4	2	9	4	13	6	18	9	22	11
500	6	3	12	6	18	9	24	12	30	15

Figure 8-5. Windage Click Chart for the Flag Method.

Once the rifle is zeroed, a change in temperature of 20 degrees or more can cause the bullet to strike above or below the point of aim. Therefore, if the temperature changes 20 degrees or more, a Marine should re-zero the rifle.

If the rifle is exposed to below freezing temperatures, it should not be brought immediately into a warm location. Condensation may form on and in the rifle, and it may freeze if re-exposed to the cold. Ice that forms inside the rifle may cause it to malfunction. Ice can form on the rear sight aperture due to condensation, making it impossible to acquire sight alignment.

Precipitation

Freezing rain and other types of precipitation may make the rifle difficult to handle, foul the rifle and cause stoppages, or build up in the barrel or compensator and cause erratic shots. Care should be taken to keep the barrel and muzzle free of water. If the rifle has been submerged, ensure the bore is drained before firing. To drain the bore, pull the charging handle slightly to the rear and hold for a few seconds while the muzzle points down. Once the barrel has been drained, turn the rifle muzzle up to allow the water to drain out of the stock.

8003. Physical and Psychological Effects of Weather on Marines

Wind

Marines can shoot effectively in windy conditions if they apply a few basic techniques and develop the proper mental attitude. The Marine can combat the wind in a number of ways:

- Make subtle changes to the basic shooting positions, such as increasing muscular tension, to reduce movement of the rifle sights.
- Select a more stable firing position.
- Seek support to stabilize the rifle.
- Hold the shot and apply the fundamentals during a lull in the wind.

Temperature

Extreme Heat

In extreme heat, a Marine may experience rapid fatigue. Heat can cause muscle cramps, heat exhaustion, heat stroke, blurred vision, and reduced concentration levels that result in inaccurate shooting. Increased fluid intake and good physical condition can offset the effects of extreme heat. Sweat running into the eyes can cause irritation and make it difficult to see the sights. Extreme heat also can create ground mirages that cause a target to appear indistinct and to drift from side to side. Heat waves or mirages may also distort the target shape or the appearance of the front sight post. A mirage created by the heat of the barrel reduces a Marine's ability to see the sight clearly. To overcome the effects of heat and accurately engage a target, a Marine should maintain a center of mass hold.

Extreme Cold

Extreme cold may affect a Marine's ability to concentrate. If a Marine's hands are numb, he will have difficulty holding a frigid rifle and executing effective trigger control. To protect the hands in a cold environment, a Marine should wear arctic mittens or gloves. To operate the rifle while wearing arctic mittens or gloves, a Marine depresses the trigger guard plunger to open the trigger guard. This allows easier access to the trigger. See figure 8-6. The hasty sling can assist in holding the hand in place on the hand guards so the hand does not slip while wearing mittens.

Precipitation

Precipitation (rain, snow, hail, sleet) can affect target engagement, a Marine's comfort level, and a Marine's ability to concentrate. The amount and type of precipitation may obscure or completely hide the target and it may reduce a Marine's ability to establish an accurate sight picture. Precipitation collecting on rear sight aperture can make it difficult to establish sight alignment and sight picture. Protect sights as much as possible

Figure 8-6. Open Trigger Guard.

during periods of precipitation. It is easy to lose concentration when wet and uncomfortable. Proper dress reduces the effects of precipitation on the Marine.

Light

Light conditions can change the appearance of a target. Light affects each Marine differently. Light can affect range estimation, visual acuity, or the placement of the tip of the front sight on the target. By maintaining a center of mass hold, the effects of light can be reduced.

Bright Light

Bright light conditions exist under a clear blue sky with no fog or haze present to filter the sunlight. Bright light can make a target appear smaller and farther away. As a result, it is easy to overestimate range. Maintaining a center of mass hold, regardless of how indistinct the target appears, ensures the best chances for an effective shot.

Overcast

An overcast condition exists when a solid layer of clouds blocks the sun. The amount of available light changes as the overcast thickens. Overcast conditions can make a target appear larger and closer. As a result, it is easy to underestimate range. During a light overcast, the target appears very distinct and the rifle sights appear very distinct, making it easy to establish sight alignment. As the overcast thickens, it becomes difficult to identify the target from its surroundings.

Haze

Hazy conditions exist when fog, dust, humidity or smoke is present. Hazy conditions can make a target appear indistinct making it difficult to establish sight picture.

CHAPTER 9. ZEROING

To be combat effective, it is essential for the Marine to know how to zero his rifle. Zeroing is adjusting the sights on the weapon to cause the shots to impact where the Marine aims. This must be done while compensating for the effects of weather and the range to the target. It is critical that Marines can zero their rifles and make the sight adjustments required to engage targets accurately.

See the appropriate technical manual for procedures on boresighting and zeroing with supplemental aiming devices (e.g., laser, night vision devices).

Note

☞The procedures in this manual are written for right-handed Marines; left-handed Marines should reverse instructions as necessary.

9001. Elements of Zeroing

There are five basic elements involved in zeroing a rifle: line of sight, aiming point, centerline of the bore, trajectory, and range. See figure 9-1.

Line of Sight

The line of sight is a straight line, which begins with the shooter's eye, proceeds through the center of the rear sight aperture, and passes across the tip of the front sight post to a point of aim on a target.

Aiming Point

The aiming point is the precise point where the tip of the front sight post is placed in relationship to target.

Centerline of the Bore

Centerline of the bore is an imaginary straight-line beginning at the chamber end of the barrel, proceeding out of the muzzle, and continuing indefinitely.

Trajectory

In flight, a bullet does not follow a straight line but travels in a curve or arc, called trajectory. Trajectory is the path a bullet travels to the target. As the bullet exits the muzzle, it travels on an upward path, intersecting the line of sight (because the sights are above the muzzle). As the bullet travels farther, it begins to drop and intersects the line of sight again.

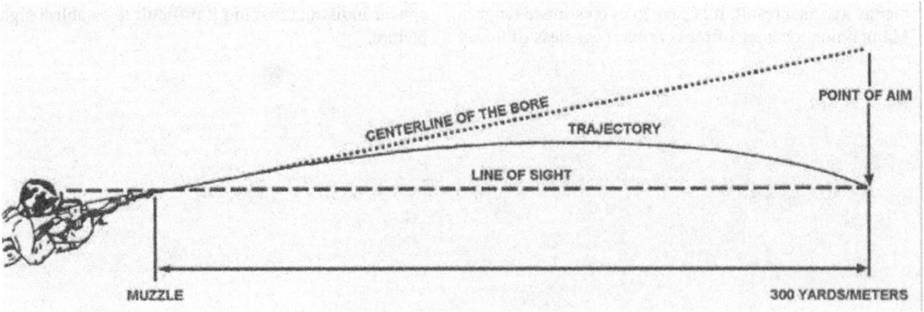

Figure 9-1. Elements of Zeroing.

Range

Range is the KD from the rifle muzzle to the target.

9002. Types of Zeros

Battlesight Zero (BZO)

A BZO is the elevation and windage settings required to place a single shot, or the center of a shot group, in a predesignated location on a target at 300 yards/meters, under ideal weather conditions (i.e., no wind). A BZO is the sight settings placed on your rifle for combat. In combat, your rifle's BZO setting will enable engagement of point targets from 0–300 yards/meters in a no-wind condition.

Zero

A zero is the elevation and windage settings required to place a single shot, or the center of a shot group, in a predesignated location on a target at a specific range, from a specific firing position, under specific weather conditions.

True Zero

A true zero is the elevation and windage settings required to place a single shot, or the center of a shot group, in a predesignated location on a target at a specific range other than 300 yards/meters, from a specific firing position, under ideal weather conditions (i.e., no wind).

9003. M16A2 Sighting System

The sighting system of the M16A2 service rifle consists of a front sight post and two rear sight apertures windage and elevation knob. Scales of the sighting system may be applied accurately to both yard and meter measurements. For example, a rear sight elevation setting of 8/3 may be used for 300 yards/meters.

Front Sight

The front sight post is used to adjust for elevation. The front sight consists of a square, rotating sight post with a four-position, spring-loaded detent (see fig. 9-2). To adjust for elevation, use a pointed instrument (or the tip of a cartridge) to depress the detent and rotate the front sight post (see fig. 9-3). To raise the strike of the bullet, rotate the post clockwise (in the direction of the arrow marked UP) or to the right. To lower the strike of the bullet, rotate the post counter-clockwise (in the opposite direction of the arrow) or to the left.

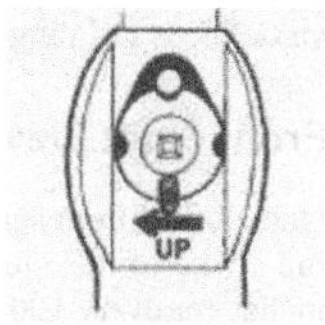

Figure 9-2. Front Sight.

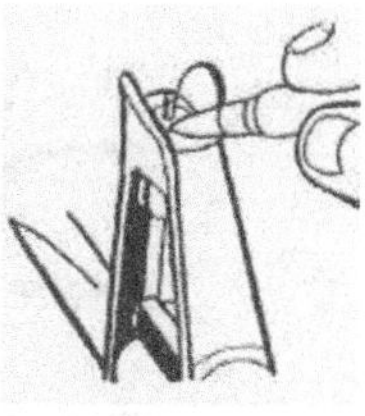
Figure 9-3. Front Sight Adjustment.

Rear Sight

The rear sight consists of two sight apertures, a windage knob, and an elevation knob. See figure 9-4. The large aperture marked 0-2 is used for target engagement during limited visibility, when a greater field of view is desired, or for engagements of targets closer than 200 yards/meters. The unmarked aperture (small aperture) is used for zeroing and normal firing situations.

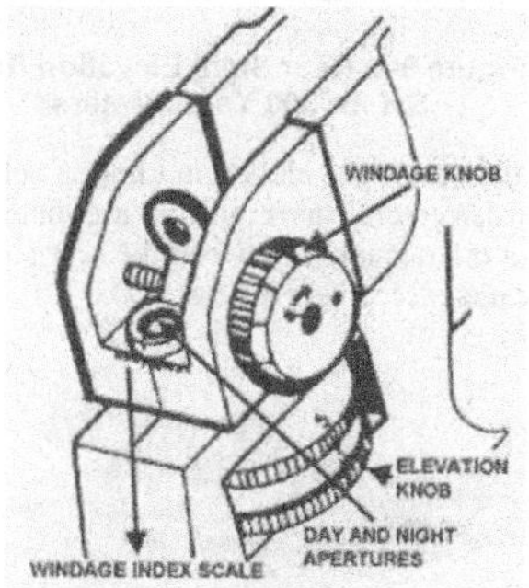

Figure 9-4. Rear Sight.

Elevation Knob

The rear sight elevation knob is used to adjust the sight for a specific range to the target. The elevation knob is indexed as shown in figure 9-5. Each number on the knob represents a distance from the target in 100-yard/meter increments. To adjust for range to the target, rotate the elevation knob so the desired setting is aligned with the index on the left side of the receiver.

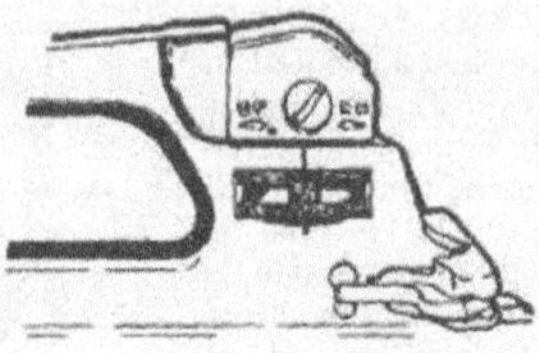

Figure 9-5. Elevation Knob.

If the elevation knob is turned so the number 8/3 aligns with the elevation index line, the 3 indicates 300 yards/meters (see fig. 9-6).

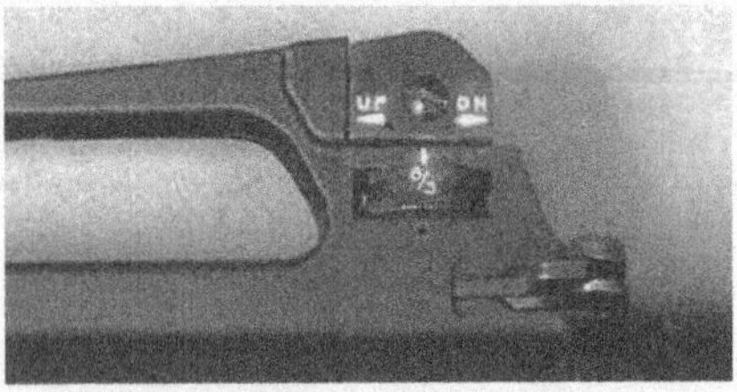

Figure 9-6. Rear Sight Elevation Knob Set for 300 Yards/Meters.

When the rear sight elevation knob is set on 8/3 for 800 yards/meters, there will be a considerable gap (about a quarter inch) between the rear sight housing and the upper receiver (see fig. 9-7).

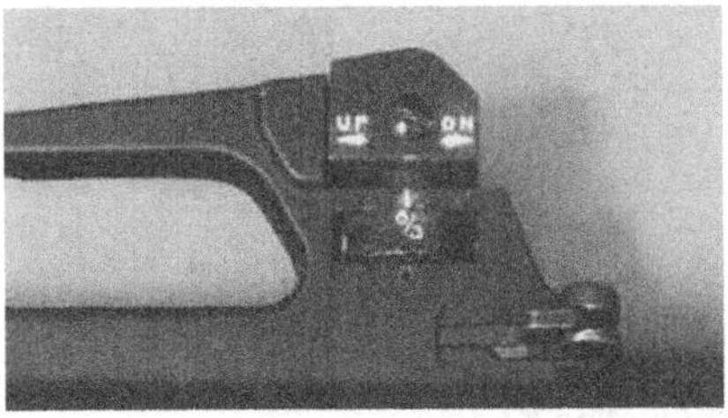

Figure 9-7. Rear Sight Elevation Knob Set for 800 Yards/Meters.

A hasty sight setting is the setting placed on the rear sight elevation knob to engage targets beyond 300 yards/meters. Hasty sight settings for ranges of 400 to 800 yards/meters are applied by rotating the rear sight elevation knob to the number that corresponds with the engagement distance of the enemy. Aligning the number 4, 5, 6 or 7 with the elevation index line places the elevation at 400, 500, 600 or 700 yards/meters, respectively. If a clockwise rotation is continued, the number 8/3 appears for the second time on the elevation index line and indicates an 800 yard/meter elevation.

Windage Knob

The windage knob is used to adjust the strike of the round right or left. The windage knob is marked with an arrow and the letter R that shows the direction the strike of the round is being moved. See figure 9-8.

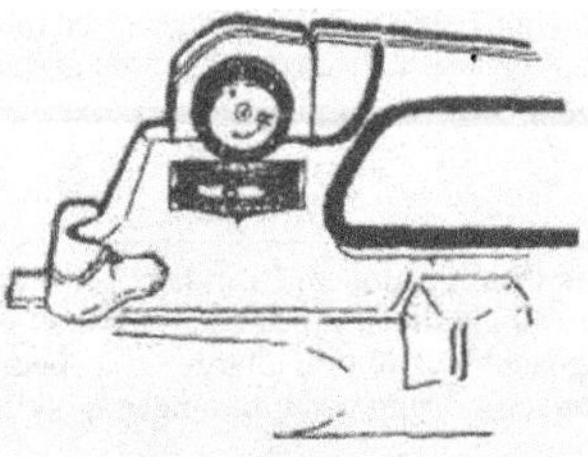

Figure 9-8. Windage Knob.

To move the strike of the round to the right, rotate the windage knob clockwise (in the direction of the arrow marked R).

To move the strike of the round to the left, rotate the windage knob counterclockwise.

9004. Windage and Elevation Rules

Moving the front sight post, elevation knob or windage knob one graduation or notch is referred to as moving one "click" on the sight. The windage and elevation rules define how far the strike of the round will move on the target for each click of front and rear sight elevation or rear sight windage for each 100 yards/meters of range to the target.

Front Sight Elevation Rule

One click of front sight elevation adjustment will move the strike of the round on target approximately 1.25 inches for every 100 yards of range to the target or 3.5 centimeters for every 100 meters of range to the target.

Rear Sight Elevation Rule

One click of rear sight elevation adjustment will move the strike of the round on the target approximately 1 inch for every 100 yards of range to the target or 2.5 centimeters for every 100 meters of range to the target.

Windage Rule

One click of windage adjustment will move the strike of the round on the target approximately 0.5 inch for every 100 yards of range to the target or 1.25 centimeters for every 100 meters of range to the target.

9005. Initial Sight Settings

Initial sight settings are those settings that serve as the starting point for initial zeroing from which all sight adjustments are made. (If the Marine already has a BZO established on his rifle, he may begin the zeroing process by using the previously established BZO sight settings.) To set the sights to initial sight settings:

Front Sight Post

To set the front sight post to initial sight setting, depress the front sight detent and rotate the front sight post until the base of the front sight post is flush with the front sight housing.

Rear Sight Elevation Knob

To set the elevation knob at the initial sight setting, perform the following:

- Rotate the rear sight elevation knob counterclockwise until the moveable rear sight housing is bottomed out on the upper receiver. See figure 9-9.

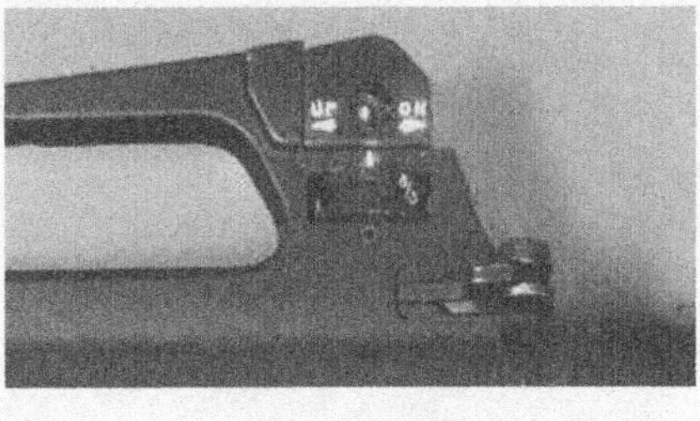

Figure 9-9. Bottoming Out Elevation Knob.

Note

Once bottomed out, the rear sight elevation knob should be three clicks counterclockwise from 8/3. If the sight fails to move three clicks counterclockwise from 8/3, it must be adjusted by a qualified armorer.

- Rotate rear sight elevation knob clockwise until the number 8/3 aligns with the index mark located on the left side of the upper receiver. See figure 9-10.

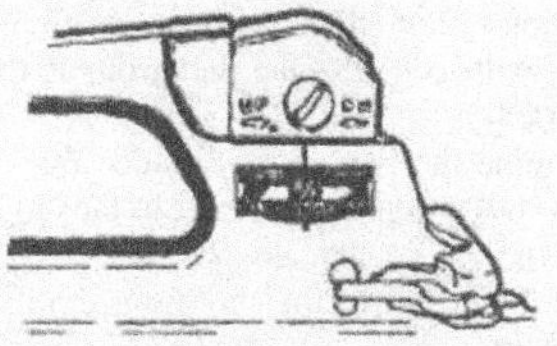

Figure 9-10. Elevation Knob Set at 8/3.

Windage Knob

To set the windage knob to initial sight setting, rotate windage knob until the index line located on the top of the large rear sight aperture aligns with the centering on the windage index scale located on the moveable base of the rear sight assembly. See figure 9-11.

Figure 9-11. Aligning Index Line.

9006. Zeroing Process

During the zeroing process, all elevation adjustments are made on the front sight post. Once a BZO is established, the front sight post should never be moved, except when rezeroing the rifle. (The rear sight elevation knob is used for dialing in the range to the target.) Zeroing is conducted at a range of 300 yards/meters. To prepare a rifle for zeroing, the rifle sights must be adjusted to the initial sight settings as outlined in paragraph 9005. Perform the following steps to zero the rifle:

- Fire a 3-shot group.
- Triangulate the shot group to find the center. See figure 9-12.
- Determine the vertical distance in inches from the center of the shot group to the center of the target.
- Make elevation adjustments on the front sight post to move the center of the shot group to the center of the target.
- Determine the horizontal distance from the center of the shot group to the center of the target.
- Make lateral adjustments on the windage knob to move the center of the shot group to the center of the target.
- Repeat preceding steps until shot group is centered.
- Fire a 4-round shot group to confirm sight setting.

Once the sight setting is confirmed, determine the value and direction of the wind and remove the number of clicks added to the windage knob (if necessary) to compensate for current wind conditions. This becomes the BZO setting for the rifle.

9007. Battlesight Zero

Zeroing is conducted at a range of 300 yards/meters. If a 300-yard/-meter range is not available, a field expedient BZO can be established at a reduced range of 36 yards/30 meters. When a rifle is zeroed for 300 yards/meters, the bullet crosses the line of sight twice. It first

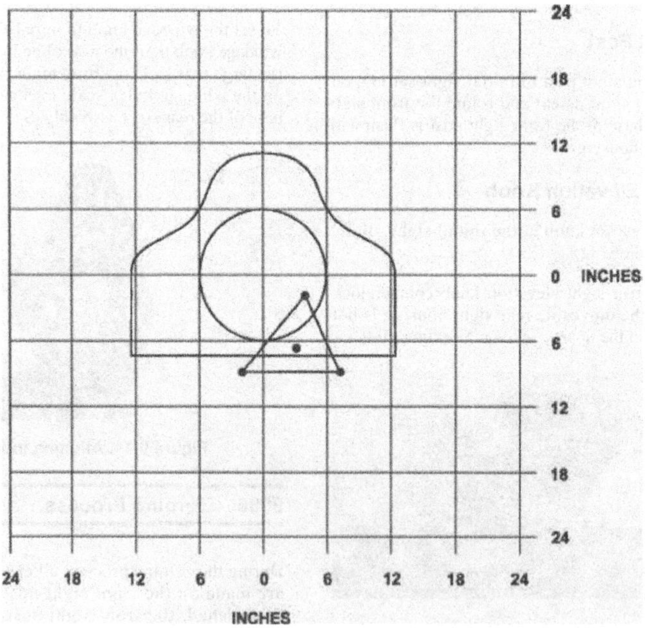

Figure 9-12. Triangulating Shot Group.

crosses the line of sight on its upward path of trajectory at 36 yards/30 meters, and again farther down range at 300 yards/meters (see fig. 9-13). Therefore a rifle's BZO may be established at a distance of 36 yards/30 meters and the same BZO will be effective at 300 yards/meters.

To establish a field expedient BZO at 36 yards or 30 meters when a 300-yard/-meter range is not available, a Marine performs the same steps as the zeroing process outlined in paragraph 9006. However, since wind does not affect the round at 36 yards/30 meters, windage is not added nor is it removed from the windage knob after confirming the BZO. To be accurate, the target must be placed exactly 36 yards (or 30 meters) from the muzzle of the rifle.

9008. Factors Causing a BZO to be Reconfirmed

Marines are responsible for maintaining a BZO on their rifles at all times. Many factors influence the BZO of a rifle. Atmospheric conditions, humidity, and temperature can cause BZOs to change on a daily basis. If operating in a combat environment, Marines should confirm their BZO as often as possible. To confirm a BZO, a Marine may begin the zeroing process by using the previously established BZO sight settings rather than placing the sights at initial sight setting. The following factors cause a BZO to be reconfirmed.

Maintenance

It is possible for the BZO to change if ordnance personnel perform maintenance on a rifle. If maintenance was performed, it is critical that the rifle be rezeroed as soon as possible.

Temperature

An extreme change in temperature (i.e., 20 degrees or more) will cause the elevation BZO to change. Changes in temperature cause chamber pressure to increase when hot and decrease when cold. This causes shots to impact the target high in hot temperatures and low in cold temperatures.

Climate

Changing climates (i.e., moving from a dry climate to a tropical climate) can mean changes in air density, moisture content, temperature or barometric pressure. Any of these elements can affect the rifle's BZO.

Ammunition

Inconsistencies in the production of ammunition lots can change a rifle's BZO.

Ground Elevation

Drastic changes in ground elevation can create changes in air density, moisture content, temperature or barometric pressure. Any of these elements can affect the rifle's BZO.

Uniform

If Marines zero their rifles in utility uniform and fire in full battle gear, their BZOs will change. The wearing of full battle gear changes eye relief, placement of the rifle in the shoulder pocket, and the way the rifle is supported on the handguard. Marines must establish their BZOs while wearing the uniform and equipment they will be wearing while engaging targets.

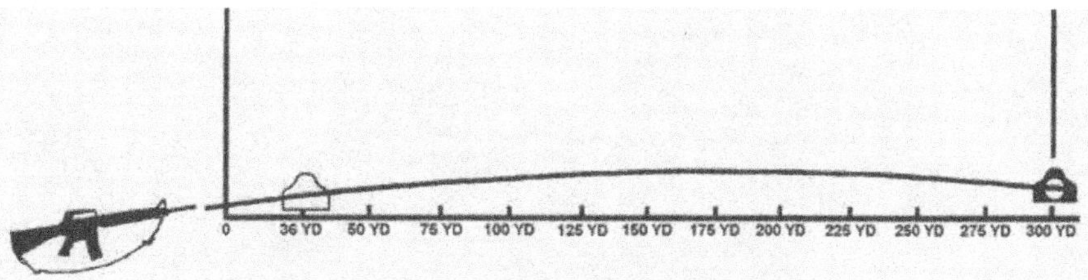

Figure 9-13. Bullet Crossing the Line of Sight Twice.

9009. Factors Affecting the Accuracy of a BZO

Anything the Marine changes from shot to shot affects the accuracy of his BZO. It further affects the accuracy of shot placement. The following factors, when applied inconsistently, diminish the accuracy of a BZO:

- Any of the seven factors (left hand, rifle butt in the pocket of the shoulder, grip of the right hand, right elbow, stock weld, breathing, and muscular relaxation/tension).
- Stability of hold.
- Sling tension.
- Trigger control.
- Sight picture.

CHAPTER 10. ENGAGEMENT TECHNIQUES

A Marine must maintain the ability to react instinctively in a combat environment—day or night. He must possess a combat mindset that eliminates hesitation, fear, or uncertainty of action. A Marine's combat mindset allows him to engage the enemy rapidly and focus on the actions required to fire well-aimed shots. He must remember that speed alone does not equate to effective target engagement. He should fire only as fast as he can fire accurately. He should never exceed his physical capability to engage a target effectively. To be effective in combat, a Marine must train to perfect the physical skills of target engagement (such as presenting the weapon and assuming a shooting position) until they become instinctive. A Marine must also employ effective engagement techniques that enable acquisition and engagement of a variety of targets in diverse combat conditions.

10001. Target Detection

To be proficient, a Marine rifleman must be able to detect targets, determine the range to targets, and accurately engage the targets. There are many variables affecting a Marine's ability to detect and determine the range to combat targets. Enemy targets on the battlefield may be single or multiple, stationary or moving, or completely hidden from view. Success in locating an enemy target will depend upon the observer's position, his skill in searching an area, and his ability to recognize target indicators.

Target Indicators

Most combat targets are detected at close range by smoke, flash, dust, noise or movement, and are usually seen only momentarily. Target indicators are anything that reveals an individual's position to the enemy. These indicators are grouped into three general areas: movement, sound, and improper camouflage.

Movement

The human eye is attracted to movement, especially sudden movement. The Marine need not be looking directly at an object to notice movement. The degree of difficulty in locating moving targets depends primarily on the speed of movement. A slowly moving target will be harder to detect than one with quick, jerky movements.

Sound

Sound can also be used to detect an enemy position. Sound may be made by movement, rattling equipment, or talking. Sound provides only a general location of the enemy, making it difficult to pinpoint a target by sound alone. However, sound can alert the Marine to the presence of a target and increase his probability of locating it through other indicators.

Improper Camouflage

There are three indicators caused by improper camouflage: shine, outline, and contrast with the background. Most targets on the battlefield are detected due to improper camouflage. However, many times an observation post or enemy firing position will blend almost perfectly with the natural background. Only through extremely careful, detailed searching will these positions be revealed.

Shine. Shine is created from reflective objects such as metal or glass. It may also come from pools of water and even the natural oils from the skin. Shine acts as a beacon to the target's position.

Outline. Most enemy soldiers will camouflage themselves, their equipment, and their positions. The outline of objects such as the body, head and shoulders, weapons, and web gear are recognizable even from a distance. The human eye will often pick up a recognizable shape and concentrate on it even if the object cannot be identified immediately. The reliability of this indicator depends upon visibility and the experience of the observer.

Contrast With the Background. Indicators in this category include objects that stand out against (contrast with) a background because of differences in color, surface, and shape. For instance, a target wearing a dark uniform would be clearly visible in an area of snow or sand. Geometric shapes, such as helmets or rifle barrels, can be easy to detect in a wooded area. Fresh soil around a fighting hole contrasts with the otherwise unbroken ground surface. While observing an area, take note of anything that looks out of place or unusual and study it in more detail. This will greatly increase your chances of spotting a hidden enemy.

Identifying Target Location

Observation Position

A good position is one that offers maximum visibility of the area while affording cover and concealment. The optimal observation position should allow the Marine to scan all the areas of observation and offer enough concealment to prevent his position from being detected.

The Marine should avoid positions that are obvious or stand out, such as a lone tree in a field or a pile of rocks on a hill. These positions may be ideal points for easy observation, but they will also make it easier for the enemy to locate the Marine.

Methods for Searching an Area

In searching an area, the Marine will be looking for target indicators. There are two techniques for searching an area: the hasty search and the detailed search.

Hasty Search. When a Marine moves into a new area, he must quickly check for enemy activity that may pose an immediate danger. This search is known as the hasty search and should take about 30 seconds, depending on the terrain.

Quickly glance at various points throughout the area rather than sweeping the eyes across the terrain in one continuous movement. The Marine should search the area nearest him first since it poses the greatest potential for danger.

This method of search is effective because it takes advantage of peripheral vision. Peripheral vision enables the detection of any movement in a wide area around the object being observed. For this technique to be effective, the eyes must be focused briefly on specific points (i.e., areas that may provide cover or concealment for the enemy).

Detailed Search. A detailed search is a systematic examination of a specific target indicator or of the entire observation area. A detailed search should be conducted immediately on target indicators located during the hasty search. The detailed search should be made from top to bottom or side to side, observing the entire object in exact detail. If multiple indicators were observed during the hasty search, the detailed search should begin with the indicator that appears to pose the greatest threat.

After a thorough search of target indicators, or if no indicators were located during the hasty search, a detailed search should be made of the entire observation area. The 50-meter overlapping strip method is normally used.

Normally, the area nearest the observer offers the greatest potential danger and should be searched first. Begin the search at one flank, systematically searching the terrain at the front in 180-degree arcs, searching everything in exacting detail, 50 meters in depth. See figure 10-1.

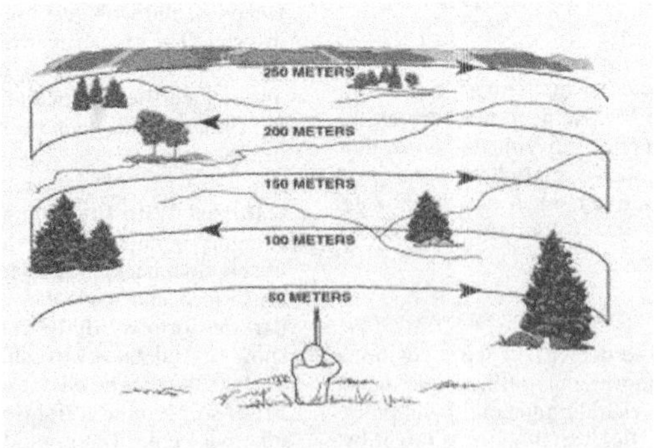

Figure 10-1. Searching the Terrain in Overlapping Strips.

After reaching the opposite flank, systematically cover the area between 40 and 90 meters from your position. The second search of the terrain includes about 10 meters of the area examined during the first search. This technique ensures complete coverage of the area.

Continue the overlapping strip search method for as far as you can see.

Maintaining Observation

Method

The combat situation will dictate the method of maintaining observation of an area. Generally, the observation method will include a combination of hasty and detailed searches.

Sequence of Observation

Observation is often conducted by a two-man team. One team member should constantly observe the entire area using the hasty search technique and the other team member should conduct a detailed overlapping strip search. If you are observing as an individual, devise a plan to ensure that the area of observation is completely covered. When entering a new area, immediately conduct a hasty search. Since a hasty search may fail to detect some indicators, periodically conduct a detailed search of the area. A detailed search should also be conducted any time your attention has been diverted from the search area.

Remembering Target Location

Most targets are seen only briefly and most areas contain multiple targets. Once you have located a target indicator, you will need to remember its location to engage it successfully. To help remember the location of a target, select a known feature and use it as a reference point to determine the distance and general direction to the target.

10002. Range Estimation

To engage targets at unknown distances, a Marine must determine the distance from his location to a known point. This is known as range estimation. The ability to determine range is a skill that must be developed if a Marine is to successfully engage targets at unknown distances. Precise range estimation enhances accuracy, enhances the chance of survival, and determines if a target can be effectively engaged using the rifle's existing BZO or if a new sight setting or point of aim is required.

Range Estimation Methods

Unit of Measure Method

To use this method, a Marine visualizes a distance of 100 meters on the ground, and then estimates how many of these units will fit between him and the target. This determines the total distance to the target. See figure 10-2.

Figure 10-2. Unit of Measure Method.

The greatest limitation of this method is that its accuracy is related to the amount of visible terrain. For example, if a target appears at a range of 500 meters or more and only a portion of the ground between a Marine and the target can be seen, it becomes difficult to use the unit of measure method to estimate range accurately. A Marine must practice this method frequently to be proficient. Whenever possible, a Marine should select an object, estimate the range, and then verify the actual range by either pacing or using another accurate measurement.

Rifle Front Sight Post Method

The area of the target covered by the rifle's front sight post can be used to estimate range to a target. A Marine notes the appearance of the front sight post on a known-distance target. A Marine then uses this as a guide to determine range over an unknown distance. Because the apparent size of the target changes as the distance to the target changes, the amount of the target covered by the front sight post varies based on the range. In addition, a Marine's eye relief and perception of the front sight post affect the amount of the target that is visible. To use this method, a Marine must apply the following guidelines:

- The front sight post covers the width of a man's chest or body at approximately 300 meters.
- If the target is less than the width of the front sight post, the target is in excess of 300 meters and the rifle's BZO cannot be used effectively.
- If the target is wider than the front sight post, the target is less than 300 meters and can be engaged point of aim/point of impact using the rifle's BZO.

See figure 10-3.

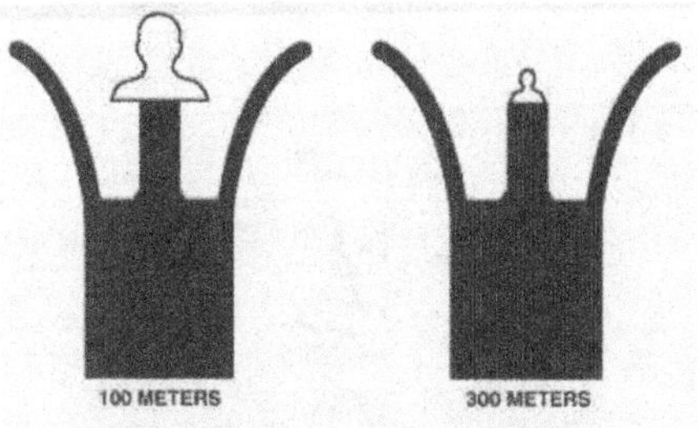

Figure 10-3. Front Sight Post Method.

Visible Detail Method

The amount of detail seen at various ranges can provide a Marine with an estimate of the target's distance. To use this method, a Marine must be familiar with the size and various details of personnel and equipment at known distances. Visibility (such as weather, smoke or darkness) limits the effectiveness of this method. A Marine should observe a man while he is standing, kneeling, and in the prone position at known ranges of 100 to 500 meters. He should note the man's size, characteristics/size of his uniform and equipment, and any other pertinent details. The Marine then uses this as a guide to determine range over an unknown distance. A Marine also should study the appearance of other familiar objects such as rifles and vehicles. To use this method, a Marine applies the following general guidelines:

- At 100 meters, the target is clearly observed in detail and facial features are distinguished
- At 200 meters, the target is clearly observed. There is a loss of facial detail. The color of the skin and equipment are still identifiable.
- At 300 meters, the target has a clear body outline, face color usually remains accurate, but remaining details are blurred.
- At 400 meters, the body outline is clear, but remaining detail is blurred.
- At 500 meters, the body shape begins to taper at the ends, and the head becomes indistinct from the shoulders.
- At 600 meters, the body appears wedge-shaped and headless.

Bracketing Method

This method of range estimation estimates the shortest possible distance and the greatest possible distance to the target. For example, a Marine estimates that a target may be as close as 300 meters but it could be as far away as 500 meters. The estimated distances are averaged to determine the estimated range to the target. For example, the average of 300 meters and 500 meters is 400 meters.

Halving Method

To use the halving method, a Marine estimates the distance halfway between him and the target, then doubles that distance to get the total distance to the target. A Marine must take care when judging the distance to the halfway point; any error made in judging the halfway distance is doubled when estimating the total distance.

Combination Method

The methods previously discussed require optimal conditions with regard to the target, terrain, and visibility in order to obtain an accurate range estimation. A Marine should estimate the range using two methods and then compare the estimates, or two Marines can compare their estimates. The average of the two estimates should be close to the actual range to target.

Factors Affecting Range Estimation

The following specific factors will affect the accuracy of estimation. A Marine must be aware of these factors and attempt to compensate for their effects.

Nature of the Target

- An object with a regular outline such as a steel helmet, rifle, or vehicle on a clear day will appear to be closer than one with an irregular outline such as a camouflaged object.

- A target that contrasts with its background will appear to be closer than a target that blends in with its background.
- A partially exposed object will appear to be farther away than it is.
- A target will appear to be farther away if the target is smaller than the objects surrounding it.

Nature of the Terrain

- Terrain that slopes upward gives the illusion of shorter distance.
- Terrain that slopes downward gives the illusion of greater distance.
- Terrain with dead space makes the target appear to be closer.
- Smooth terrain such as sand, water, or snow gives the illusion of greater distance.

Limited Visibility

The position of the light source significantly affects a Marine's ability to estimate range. Other factors that affect range estimation include smoke, fog, rain, angled light, or anything that obscures the battlefield. Generally, when the sun is bright, a target appears further away. When the sky is overcast, a target generally appears closer. Target contrast is another factor to consider when estimating range as well as obstacles located between the shooter and the target. If there is contrast (e.g., color variation) between the target and the background, the target will appear closer. If there is little or no contrast between the target and the background, the target will appear farther away. If there is an object between the target and the shooter that will distract the shooter (e.g., a tree), the target will appear farther away.

10003. Offset Aiming

The conditions of rifle fire in combat may not permit mechanical adjustments of the sights. To engage a target during combat, a Marine may be required to aim his rifle at a point on the target other than center mass. This is known as offset aiming. Offset aiming is used to compensate for the distance and size of the target, wind, and speed and angle of a moving target. There are two primary techniques for offset aiming: point of aim technique and known strike of the round.

Point of Aim Technique

The point of aim technique is the shifting of the point of aim (sight picture) to a predetermined location on or off the target to compensate for a known condition (i.e., wind, distance, and movement). Each predetermined location is known as a point of aim. Figure 10-4 illustrates points of aim for elevation.

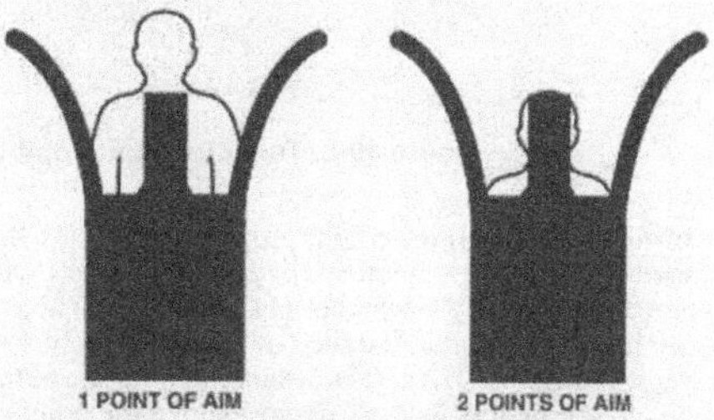

Figure 10-4. Points of Aim for Elevation.

Elevation

Predetermined points of aim sector the target horizontally. The tip of the front sight post held at shoulder level is considered one point of aim; the tip of the front sight post held at the top of the target's head is considered two points of aim. A Marine uses these points of aim to compensate for the elevation required to engage a target beyond the BZO capability of the rifle or to engage a small target (e.g., head shot) inside the BZO of the weapon.

Beyond the BZO. To use the point of aim technique to engage a target beyond the BZO of the rifle, a Marine must apply the following guidelines:

- When range to the target is estimated to be beyond 300 meters out to 400 meters, hold one point of aim.
- When the range to the target is estimated to be beyond 400 meters out to 500 meters, hold two points of aim.

Note

It is better to apply a hasty sight setting at ranges beyond the rifle's BZO. Points of aim are only guidelines at these distances because the front sight will mask the target when the front sight is held above center mass, making it difficult to acquire sight picture.

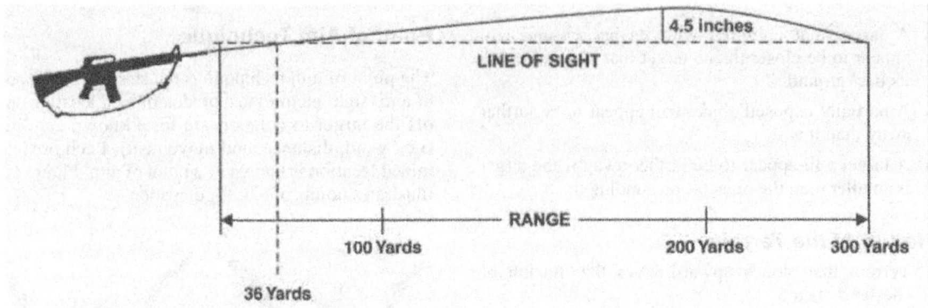

Figure 10-5. Trajectory and Point of Aim/Point of Impact for 300 Yard BZO.

Inside the BZO. If the rifle is properly zeroed for 300 yards/meters, the trajectory (path of the bullet) will rise approximately 4.5 inches (11 centimeters) above the line of sight at a distance of approximately 175 yards (160 meters). At other distances, the strike of the bullet will be less than 4.5 inches above the point of aim. Only at 36 yards/30 meters and 300 yards/meters does the point of impact coincide with the point of aim. If only a portion of the target is visible (e.g., the head of an enemy soldier), the trajectory of the bullet may have to be taken into consideration when firing at a distance less than 300 yards/meters. If a Marine does not consider trajectory, he may shoot over the top of the target if the target is small and at a distance less than 300 yards/meters. See figure 10-5.

Windage

Predetermined points of aim sector the target vertically (see fig. 10-6). The tip of the front sight post centered on the edge of the target into the wind is considered one point of aim; the trailing edge of the front sight post held on the edge of the target into the wind is considered two points of aim. The same units of measure are applied off the target for holds of additional points of aim. These points of aim are used to compensate for wind affecting the strike of the round and when there is no time to adjust the rifle's sights, or when a lead is required to engage a moving target (points of aim for moving targets are discussed in paragraph 10007). Table 10-1 provides points of aim for full value winds.

Known Strike of the Round

This offset aiming technique shifts the aiming point (sight picture) to compensate for rounds that strike off target center. The known strike of the round method is used if the strike of the round is known. To engage a target using this method, a Marine aims an equal distance from center mass opposite the known strike of the round. For example, if the round strikes high and left, a Marine aims an equal and opposite distance low and right.

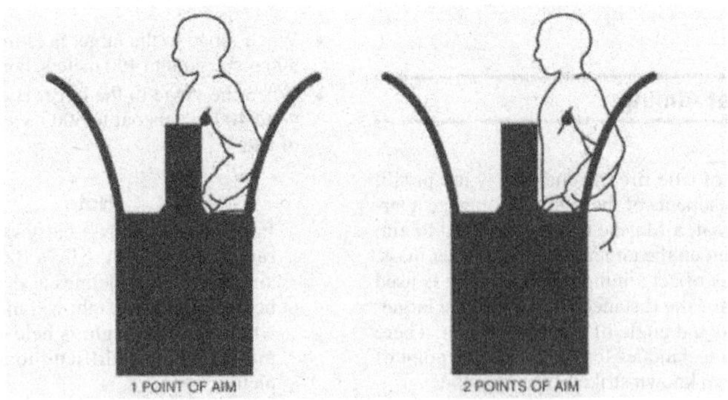

Figure 10-6. Points of Aim.

Table 10-1. Points of Aim for Full Value Winds.

Wind	Distance (yards/meters)	Points of Aim
Light	0-300	0
Medium	200-300	1
Strong	0-200	1
Strong	200-300	2

10004. Techniques of Fire

The size and distance to the target should dictate the technique of fire.

Two-Shot Technique

In combat, an effective technique for eliminating a threat is to rapidly fire more than one shot on the target. Two shots fired in rapid succession will increase the trauma (i.e., shock, blood loss) on the target, increasing the Marine's chances of quickly eliminating the threat. Firing two shots enables the Marine to break out of the tunnel vision often associated with firing in combat and then assess the situation to determine follow-on action. To execute the two-shot technique, the Marine acquires sight picture for each shot fired (each pull of the trigger).

Single Shot Technique

If the target is at a long range or if the target is small (i.e., partially exposed), it can best be engaged with a single, precision shot. A Marine's stability of hold and sight alignment are more critical to accurate engagement of long-range or small targets. To engage a target with the single shot technique, the Marine must slow down the application of the fundamentals and place one well-aimed shot on target.

Sustained Rate of Fire

An effective method for delivering suppressive fire is to fire at the sustained rate of 12 to 15 rounds per minute. Management of recoil is critical to bring the sights back on target after the shot is fired.

Three-Round Burst Technique

When set on burst, the design of the rifle permits three shots to be fired from a single trigger pull. The rounds fire as fast as the weapon will function and cause the muzzle to climb during recoil. The ability to manage recoil is extremely important when firing the rifle on burst. To achieve the desired effect (i.e., 3 rounds on target), the Marine must control the jump angle of the weapon to maintain the sights on target. At short ranges (i.e., 25 meters or less), firing on three-round burst can be an effective technique to place rounds on a man-sized target quickly to increase trauma on the target. To execute the three-round burst technique, the Marine places the selector lever on burst, aims center mass, and acquires sight picture once for the single trigger pull.

10005. Engaging Immediate Threat Targets

Immediate threat target engagement is characterized by short-range engagement (i.e., less than 50 meters) with little or no warning that requires an immediate response to engage an enemy. This type of engagement is likely in close terrain (e.g., urban, jungle). If this type of engagement is likely, the large rear sight aperture (0-2) could be placed up to provide a wider field of view and detection of targets. Marksmanship skills include quick presentation and compression of the fundamentals (i.e., quick acquisition of sight picture, uninterrupted trigger control). At close ranges, perfect sight alignment is not as critical to accuracy on target. However, the front sight post must be in the rear sight aperture; proper sight alignment is always the goal.

10006. Engaging Multiple Targets

When engaging multiple targets, a Marine must prioritize each target and carefully plan his shots to ensure successful target engagement. Mental preparedness and the ability to make split-second decisions are the key to successful engagement of multiple targets. The proper mindset allows a Marine to react instinctively and to control the pace of the battle rather than just reacting to the threat.

After the first target is engaged, a Marine must immediately engage the next target and continue to engage targets until they are eliminated. While engaging multiple targets, a Marine must be aware of his surroundings and not fixate on just one target. He must rapidly prioritize the targets, establish an engagement sequence, and engage the targets. A Marine also must maintain constant awareness and continuously search the terrain for additional targets.

Prioritizing Targets

The combat situation will usually dictate the order of multiple target engagement. Target priority is based on factors such as proximity, threat, and opportunity; no two situations will be the same. The principal method is to determine the level of threat for each target so all may be engaged in succession from the most threatening to the least threatening. The target that poses the greatest threat (e.g., closest, greatest firepower) should be engaged first. Prioritizing targets is an ongoing process. Changes in threat level, proximity, or the target itself may cause a Marine to revise his priorities. Therefore, a Marine must remain alert to changes in a target's threat level and proximity and other target opportunities as the battle progresses.

Technique of Engagement

To engage multiple targets, the Marine performs the following steps:

- The first target with two rounds.
- The recoil of the rifle can be used to direct the recovery of the weapon on to the next target. As the weapon is coming down in its recovery, the Marine physically brings the sights onto the desired target. Pressure is maintained on the trigger throughout recovery and trigger control is applied at a rate consistent with the Marine's ability to establish sight picture on the desired target.
- When possible, such as when all targets are of equal threat, the Marine should engage targets in a direction that maximizes his ability to support and control the weapon.
- The preceding steps are repeated until all targets have been engaged.

Firing Position

The selection and effective use of a firing position are critical to the successful engagement of multiple targets. A Marine should make a quick observation of the terrain and select a firing position that provides good cover and concealment, as well as the flexibility to engage multiple targets. If enemy targets are widely dispersed, the selected position must provide the Marine with flexibility of movement. The more restrictive the firing position, the longer it will take a Marine to eliminate multiple targets.

Prone

The prone position limits left and right lateral movement and is, therefore, not recommended for engaging short-range dispersed targets. Because the elbows are firmly placed on the ground in the prone position, upper body movement is restricted.

Sitting

Like the prone position, the sitting position allows limited lateral movement. This makes engagement of widely-dispersed multiple targets difficult. To ease engagement, the forward arm can be moved by pivoting on the elbow, but this movement disturbs the stability of the position.

Kneeling

The kneeling position provides a wider, lateral range of motion since only one elbow is used for support. A Marine moves from one target to another by rotating at the waist to move the forward arm in the direction of the target, either right or left.

Standing

The standing position allows maximum lateral movement. Multiple targets are engaged by rotating the upper body to a position where the sights can be aligned on the desired target. If severe or radical adjustments are required to engage widely dispersed targets, a Marine moves his feet to establish a new position rather than give up maximum stability of the rifle. This avoids poorly placed shots that can result from an unstable position.

10007. Engaging Moving Targets

In combat, it is unlikely that a target will remain stationary. The enemy will move quickly from cover to cover, exposing himself for the shortest possible time. Therefore, a Marine must quickly engage a moving target before it disappears.

Types of Moving Targets

There are two types of moving targets: steady moving targets and stop and go targets. Steady moving targets move in a consistent manner and remain in a Marine's field of vision. A walking or running man is an example of a steady moving target. A stop and go target appears and disappears during its movement. A stop and go target will present itself for only a short time before it reestablishes cover. An enemy moving from cover to cover is an example of a stop and go target. This target is most vulnerable to fire at the beginning and end of its movement to new cover because the target must gain momentum to exit its existing cover and then slow down to occupy a new position.

Leads

When a shot is fired at a moving target, the target continues to move during the time the bullet is in flight. Therefore, a Marine must aim in front of the target; otherwise, the shot will fall behind the target. This is called leading a target. Lead is the distance in advance of the target that the rifle sights are placed to accurately engage the target when it is moving.

Amount of Lead Required

Factors that affect the amount of lead are the target's range, speed, and angle of movement.

Range. Lead is determined by the rifle's distance to the target. When a shot is fired at a moving target, the target continues to move during the time the bullet is in flight. This time of flight could allow a target to move out of the bullet's path if the round was fired directly at the target. Time of flight increases as range to the target increases. Therefore, the lead must be increased as the distance to the target increases.

Speed. If a man is running, a greater lead is required because the man will move a greater distance than a walking man will while the bullet is in flight.

Angle of Movement. The angle of movement across the line of sight relative to the flight of the bullet determines the type (amount) of lead.

Three Types of Leads: Full, Half, No

- **Full Lead**. The target is moving straight across a Marine's line of sight with only one arm and half the body visible. This target requires a full lead because it will move the greatest distance across a Marine's line of sight during the flight of the bullet.
- **Half Lead**. The target is moving obliquely across a Marine's line of sight (at a 45-degree angle). One arm and over half the back or chest are visible. This target requires half of a full lead because it will move half as far as a target moving directly across a Marine's line of sight during the flight of the bullet.
- **No Lead**. The target is moving directly toward or away from a Marine and presents a full view of both arms and the entire back or chest. No lead is required. A Marine engages this target as if it were a stationary target because it is not moving across his line of sight.

Point of Aim Technique

Predetermined points of aim sector the target vertically. The tip of the front sight post centered on the leading edge of the target is considered one point of aim; the trailing edge of the front sight post held on the leading edge of the target is considered two points of aim. The same units of measure are applied off the target for holds of additional points of aim. These points of aim are used to compensate when a lead is required to engage a moving target. The following guidelines apply if a Marine uses the point of aim technique to establish a lead for a moving target at various ranges and speeds (see fig. 10-7, on page 10-10). These guidelines do not consider wind or other effects of weather. Body width in these examples is considered to be 12 inches (30 centimeters) (side view of the target).

For a slow walking target (approximately 2 to 2.5 mph/3.2 to 4 kph) moving directly across the line of sight (full lead)—

- At a range of 200 meters or less, no lead is required.
- At a range of 300 meters, hold one point of aim in the direction the target is moving.

For a fast walking target (approximately 4 mph/6.4 kph) moving directly across the line of sight (full lead)—

- At a range of 200 meters or less, hold one point of aim in the direction the target is moving.
- At a range of 300 meters, hold two points of aim in the direction the target is moving.

For a target running (approximately 6 mph/9.7 kph) directly across the line of sight (full lead)—

- At a range of 100 meters or less, hold one point of aim in the direction the target is moving.

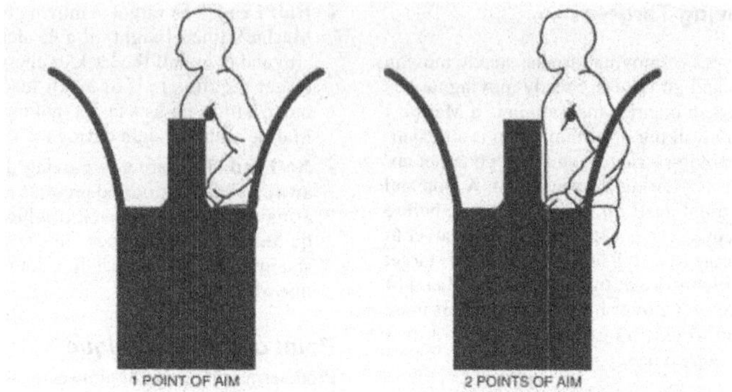

Figure 10-7. Points of Aim.

- At a range of 200 meters, hold two points of aim in the direction the target is moving.

For a target moving at a 45-degree angle (an oblique target) across the line of sight, the lead is one half of the lead that is required for a target moving directly across the line of sight.

Engagement Methods

Moving targets are the most difficult targets to engage. However, they can be engaged successfully by using the tracking or the ambush method.

The Tracking Method

The tracking method is used for a target that is moving at a steady pace over a well-determined route. If a Marine uses the tracking method, he tracks the target with the rifle's front sight post while maintaining sight alignment and a point of aim on or ahead of (leading) the target until the shot is fired. When establishing a lead on a moving target, rifle sights will not be centered on the target and instead will be held on a lead in front of the target. See figure 10-8. A Marine performs the following steps to execute the tracking methods:

- Present the rifle to the target.
- Swing rifle muzzle through the target (from the rear of the target to the front) to the desired lead (point of aim). The point of aim may be on the target or some point in front of the target depending upon the target's range, speed, and angle of movement.
- Track and maintain focus on the front sight post while acquiring the desired sight picture. It may be necessary to shift the focus between the front sight post and the target while acquiring sight picture, but the focus must be on the tip of the front sight post when the shot is fired.
- Engage the target once sight picture is acquired while maintaining the proper lead.
- Follow-through so the lead is maintained as the bullet exits the muzzle.
- Continue to track in case a second shot needs to be fired on the target.

The Ambush Method

The ambush method is used when it is difficult to track the target with the rifle, as in the prone, sitting, or any supported position. The lead required to effectively engaging the target determines the engagement point. With the sights settled, the target moves into the predetermined engagement point and creates the desired sight picture. See figure 10-9.

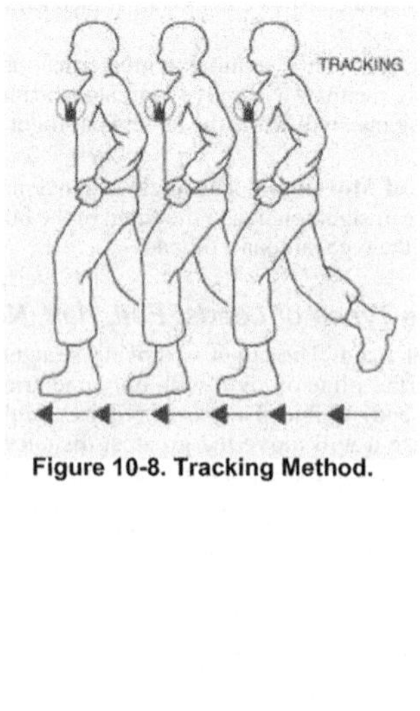

Figure 10-8. Tracking Method.

The trigger is pulled simultaneously with the establishment of sight picture. To execute the ambush method, a Marine performs the following steps:

- Select an aiming point ahead of the target.
- Obtain sight alignment on the aiming point.
- Hold sight alignment until the target moves into vision and the desired sight picture is established.
- Engage the target once sight picture is acquired.
- Follow-through so the rifle sights are not disturbed as the bullet exits the muzzle.

A variation of the ambush method can be used when engaging a stop and go target. A Marine should look for a pattern of exposure (e.g., every 15 seconds). Once a pattern is determined, a Marine establishes a lead by aiming at a point in front of the area in which the target is expected to appear, then he fires the shot at the moment the target appears.

Marksmanship Fundamentals

Engaging moving targets requires concentration and adherence to the fundamentals of marksmanship. The following modifications to the fundamentals of marksmanship are critical to engagement of moving targets.

Sight Picture

Typically, sight picture is the target's center of mass. If a Marine engages a moving target, he bases his sight picture on the target's range, speed, and angle of movement, i.e., sight alignment may be established on a point of aim in front of the target.

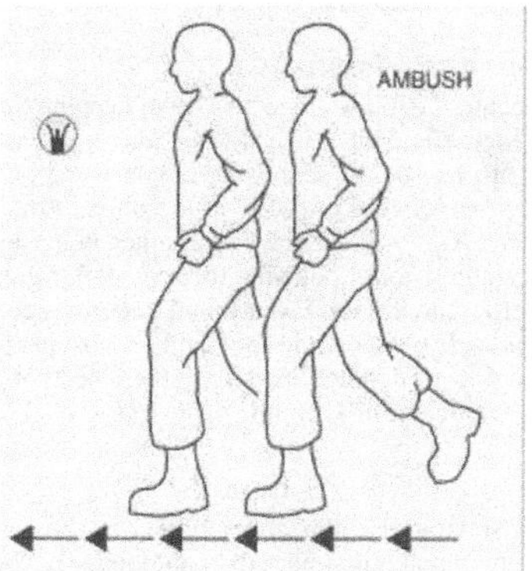

Figure 10-9. Ambush Method.

Trigger Control

As with any target engagement, trigger control is critical to the execution of shots. A Marine can apply pressure on the trigger prior to establishing sight picture, but there should be no rearward movement of the trigger until sight picture is established. Interrupted trigger control is not recommended because the lead will be lost or have to be adjusted to reassume proper sight picture. When using the tracking method, continue tracking as trigger control is applied to ensure the shot does not impact behind the moving target.

Follow-through

If a Marine uses the tracking method to engage moving targets, he continues to track the target during follow-through so the desired lead is maintained as the bullet exits the muzzle. Continuous tracking also enables a second shot to be fired on target if necessary.

Stable Position

To engage moving targets using the tracking method, the rifle must be moved smoothly and steadily as the target moves. A stable position steadies the rifle sights while tracking. Additional rearward pressure may be applied to the pistol grip to help steady the rifle during tracking and trigger control. Elbows may be moved from the support so the target can be tracked smoothly.

10008. Engaging Targets at Unknown Distances

Hasty Sight Setting

While a BZO is considered true for 300 meters, a Marine must be capable of engaging targets beyond this distance. The rifle's sighting system allows sight settings for distances out to 800 meters in 100-meter increments. If a Marine must establish a BZO for extended ranges, it is referred to a hasty sight setting. To achieve a hasty sight setting, a Marine dials the appropriate range numeral on the rear sight elevation knob that corresponds to the range to the target. For example, if the rear sight elevation knob is set at 8/3 and a target appears at 500 meters, rotate the knob to the 5 setting.

Note

Upon completion of firing with a hasty sight setting for extended ranges, return the rear sight to the BZO setting.

Point of Aim Technique

If the distance to the target is beyond the BZO capability of the rifle and time does not permit adjustment of the sights, a Marine can use offset aiming techniques to apply a point of aim for elevation to engage the target. See paragraph 10003.

10009. Engaging Targets During Low Light and Darkness

Combat targets are frequently engaged during periods of darkness or under low-light conditions. Although basic marksmanship fundamentals do not change, the principles of night vision must be applied and target detection is applied differently. During periods of darkness or low light, a Marine's vision is extremely limited. A Marine must apply the techniques of night observation in order to detect potential targets, and he must develop skills that allow him to engage targets under these conditions.

Night Vision

A Marine can improve his ability to see during periods of darkness or low light by obtaining and maintaining night vision. Since adapting to night vision is a slow and gradual process, steps should be taken to protect night vision once it is obtained.

Obtaining Night Vision

There are two methods used to obtain night vision. The first method is to remain in an area of darkness for about 30 minutes. This area can be indoors or outdoors. The major disadvantage of this approach is that an individual is not able to perform any tasks while acquiring night vision in total darkness. The second method is to remain in a darkened area under low intensity red light (similar to the light used in a photographer's darkroom) for about 20 minutes, followed by about 10 minutes in darkness without the red light. This method produces almost complete night vision adaptation while permitting the performance of some tasks during the adjustment period.

Maintaining Night Vision

Because the eyes take a long time to adjust to darkness, it is important to protect night vision once it is acquired. To maintain night vision—

- Avoid looking at any bright light. Bright light will eliminate night vision and require readaptation.
- Shield eyes from parachute flares, from spotlights or from headlights.
- When using a flashlight to read a map or any other written material:
  - Put one hand over the glass to limit the area illuminated and the intensity of the light. Keeping one eye shut will reduce the amount of night vision lost.
  - Cover the light with a red filter to help reduce the loss of night vision.
  - Minimize the time spent using a flashlight.

Factors Affecting Night Vision

Some physical factors may affect your night vision and reduce your ability to see as clearly as possible in low light or darkness. These factors include—

- Fatigue.
- Lack of oxygen.
- Long exposure to sunlight.
- Heavy smoking.
- Drugs.
- Headaches.
- Illness.
- Consumption of alcohol within the past 48 hours.
- Improper diet.

Searching Methods

Once night vision has been acquired, the Marine can locate targets. Some daylight observation techniques (e.g., searching for target indicators) also apply during periods of darkness or low light.

Off-center Vision

Off-center vision is the technique of keeping the attention focused on an object without looking directly at it (see fig. 10-10). To search for targets using off-center vision, never look directly at the object you are observing. You will see the object much better by using off-center vision. Look slightly to the left, right, above or below the object. Experiment and practice to find the best off-center angle for you. For most people, it is about 6 to 10 degrees away from the object, or about a fist's width at arm's length.

Note

Staring at a stationary object in the dark may make it appear to be moving. This occurs because the eye has nothing on which

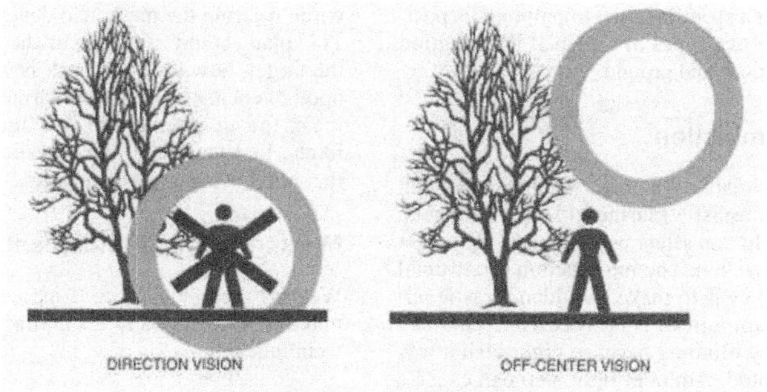

Figure 10-10. Off-center Vision.

to reference the exact position of the object. This illusion can be prevented by visually aligning the object against something else, such as a finger at arm's length.

Scanning/Figure Eight Scan

Scanning is the use of off-center vision to observe an area or object and involves moving the eyes in a series of separate movements across the objective area.

A common method of scanning is to move the eyes in a figure eight pattern (see fig. 10-11). A Marine moves the eyes in short, abrupt, irregular movements over and around the area. Once a target indicator is detected, focus is concentrated in that area, but not directly at it. Pause a few seconds at each point of observation

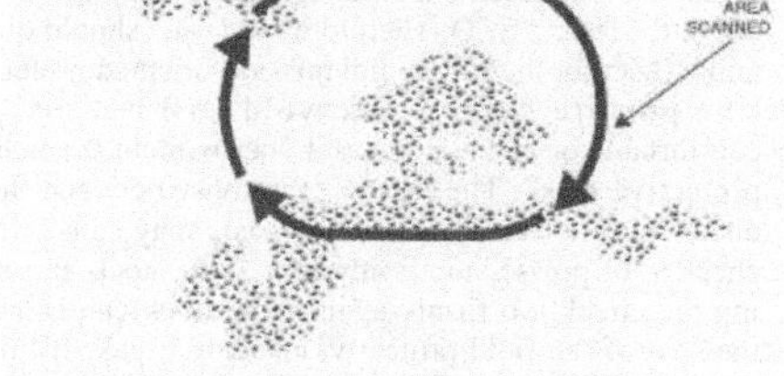

Figure 10-11. Figure Eight Scan.

since the eyes cannot focus on a still object while in motion. Rest the eyes frequently when scanning.

While you are observing, there may be periodic blackouts of night vision due to simple fatigue. This is normal and is not a cause for alarm. Night vision will quickly return after the eyes are moved and blinked a few times.

It is more effective to scan from a prone position or a position closer to the ground than the object being observed. This creates a silhouetted view of the object.

When scanning an area, look and listen for the same target indicators as in daylight: movement, sound, and improper camouflage.

Objects in bright moonlight/starlight cast shadows just as in sunlight.

Sound seems louder at night than during daylight.

Types of Illumination

There are two types of illumination that assist engagement during low light or darkness: ambient light and artificial illumination. Both ambient light and artificial illumination can affect target perception (distance and size) and night vision capabilities.

- Ambient light is the light produced by natural means (i.e., the sun, moon, and stars). Variations occur in ambient light due to the time of day, time of year, weather conditions, terrain, and vegetation.
- Artificial illumination is the light produced by a process other than natural means. Artificial light can be used to illuminate an area for target detection

or to illuminate a specific target to pinpoint its position. There are two types of artificial illumination used in combat: air and ground.

Effects of Illumination

In some combat situations, ambient light and artificial illumination may assist a Marine in locating targets. However, this light can affect perception of the target and disrupt night vision. The introduction of artificial light requires the eyes to make a sudden, drastic adjustment to the amount of light received. This can cause a temporary blinding because night vision was abruptly interrupted. Ambient light also can cause a blinding effect; e.g., a Marine may experience temporary blindness or reduced night vision if a bright moon suddenly appears from behind the clouds.

- Light behind a Marine or light between the Marine and a target illuminates the front of the target and makes it appear closer than it is.
- Light beyond the target displays the target in silhouette and makes it appear farther away than it is. If the target is silhouetted, it is easier to see and easier to engage.
- Air illumination devices are in constant motion as they descend to the ground. This movement creates changing shadows on any illuminated target causing a stationary target to appear as if it is moving.

10010. Engaging Targets while Wearing the Field Protective Mask

While engaging targets in a combat environment, a Marine is under considerable stress caused by fear, fatigue, and the noise of battle. His stress is further aggravated by the fear and uncertainty associated with a nuclear, biological, and chemical (NBC) threat. However, a Marine must be able to operate under any battlefield condition, including an NBC environment. If a Marine wears the field protective mask, its bulk and reduced visibility can affect his firing position which in turn affects the rifle's zero and his ability to engage the target. A Marine must make adjustments to his firing position and the application of marksmanship fundamentals to counter the additional gear worn in an NBC environment. Therefore, a Marine should practice wearing his field protective mask when he is not in a combat environment. This allows him to overcome any restrictions caused by the mask, develop confidence in his ability to execute well-aimed shots while wearing the mask, and develop a plan of action. This plan should address how the rifle is presented to the target, how long the mask is worn, and the likelihood of enemy contact. If a Marine expects to wear the mask for an extended period and enemy contact is likely, he should consider adjusting the rifle sights so that his first rounds are on target.

Marksmanship Fundamentals

Wearing the field protective mask requires a Marine to make modifications to his aiming and breath control techniques.

Aiming

Wearing the field protective mask affects the aiming process and the ability to locate targets. The bulk of the mask may require an adjustment to stock weld, eye relief, head position, and placement of the buttstock in the shoulder.

Breath Control

Wearing the field protective mask affects breath control because breathing may be more difficult. Temporary fogging of the lens also may be experienced. If fogging occurs, a Marine should take a deep breath and fire while holding a full breath of air (inhaling clears the fog).

Firing Position

A good firing position provides balance, control, and stability during firing. The field protective mask's added bulk and other restrictions may require a Marine to make changes to his firing position. The adjustments are unique to each Marine and based on his body size and shape and his ability to adapt to the mask. Adjustments should be minor. However, all firing positions will be affected in the following areas:

Stock Weld

Changing the placement of the cheek on the stock may affect the rifle's BZO. Therefore, a Marine should obtain a BZO for the rifle in full mission-oriented protective posture gear. Stock weld will not be as comfortable or feel as solid as it does without the field protective mask. The loss of sensitivity between the cheek and the stock, due to the mask, may cause the cheek to be pressed too firmly against the stock. Pressing the cheek too firmly against the stock can cause the seal of the field protective mask to break. If this occurs, quickly clear the mask and resume a firing position. If the lens of the field protective mask fogs up while in a firing position, this indicates that the mask's seal has been broken. Clear the mask and resume the firing position.

Eye Relief

The added bulk of the field protective mask may increase eye relief because the head is farther back along the stock. If the eye is too far from the rear sight aperture it may be difficult to acquire the target and to maintain sight picture; if the eye is too close, the rear sight can hit the mask, possibly breaking a lens or its seal.

Head Position

The mask's shape and bulk can make sight alignment difficult to achieve. The restrictive vision caused by the mask may force a Marine to roll or tilt his head over the stock to achieve sight alignment. The Marine should keep his head as erect as possible while maintaining sight alignment.

Placement of Buttstock in the Shoulder

Placement of the buttstock in the shoulder pocket may have to be altered due to the mask's added bulk. If the rifle is canted, a Marine may place the buttstock of the rifle just outside the pocket of the shoulder to achieve sight alignment. Holding the rifle straight is the preferred method of obtaining sight alignment. However, if sight alignment cannot be achieved in this position, a Marine may alter the hold of the rifle to bring the aiming eye in line with the sights. Canting the rifle drastically affects the rifle's zero. A Marine should cant the rifle only as much as is needed to obtain a good stock weld and proper sight alignment. If the rifle is canted, the point of impact may not coincide with the point of aim. For example, when wearing the mask, a right-handed Marine's point of impact is usually high and to the left of center mass (for a left-handed Marine, high and to the right of center mass). Therefore a Marine has to offset aim an equal and opposite distance low and to the right. See paragraph 10003 for a discussion on offset aiming and the known strike of the round technique.

Appendix A

DATA BOOK

Note

The principles for recording data in the data book are the same for the Entry Level Rifle (ELR) Program and the Sustainment Level Rifle (SLR) Program. This appendix provides generic information for completing the data book. Specifics pertaining to each program are contained in the ELR and SLR lesson plans.

1. Data Book

Of all the tools that assist the Marine in firing accurately and consistently, the data book, if properly used, is the most valuable asset. It contains a complete record of every shot fired and the weather conditions and their effects on shooting. When used properly it will assist the Marine in establishing and maintaining a battlesight zero (BZO).

2. Recording Data Before Firing

Recording information in the data book prior to firing saves valuable time on the firing line that should be used to prepare for firing. Some information can be recorded before going to the firing line. In the BEFORE FIRING section of the data book, record the following (see fig. A-1):

True Zero

Front Elevation

Enter the front sight post setting by recording the number of clicks up (+) or down (-) under FRONT ELEV.

Rear Elevation

Circle the rear sight elevation knob setting under REAR ELEV for the yard line firing from—

- 200 yards: 8/3-2
- 300 yards: 8/3
- 500 yards: 5

Note

At 500 yards, the rear sight elevation knob setting may be plus or minus one or two clicks off of 5.

Wind

Under the WIND column under TRUE ZERO, the R represents clicks right on the rifle from the initial sight setting and the L represents clicks left on the rifle. Enter the rear sight windage knob setting by recording the number of clicks right (clockwise) or left (counterclockwise) under WIND.

Wind

Prior to firing, check the wind. If wind conditions are present, a sight adjustment will have to be made prior to firing to ensure shots impact the center of the target.

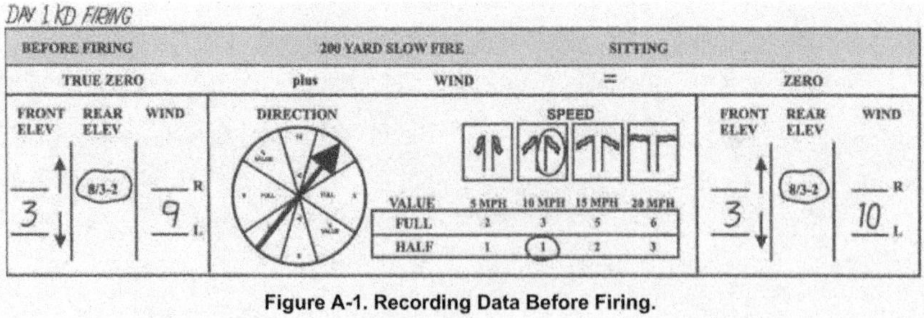

Figure A-1. Recording Data Before Firing.

Direction

Determine the direction of the wind and draw an arrow through the clock indicating the direction the wind is blowing. If the wind is not blowing, a data entry is not needed.

Note

Remember that your position is represented in the center of the clock and the target is at 12 o'clock.

Value

Look at the clock to determine if the wind is full, half, or no value wind. Under VALUE, circle FULL or HALF to indicate the wind value.

Speed

Observe the flag on the range and circle the appropriate flag indicating the wind's velocity (SPEED).

Determining Windage Adjustment

The chart beneath the flag indicates the number of clicks on the rear sight windage knob to offset the effects of the wind. Circle the number of clicks where the wind value and wind speed intersect.

Zero

Determine the zero you will place on your rifle to accommodate wind conditions to begin firing. This ZERO will be the TRUE ZERO plus the rear sight windage setting to compensate for the effects of wind.

Front Elev and Rear Elevation

Elevation adjustments are not affected by wind so the same settings are carried over from TRUE ZERO.

Wind

Wind will affect the strike of the round right or left on the target. Therefore, if wind is a factor, the rear sight windage knob must be adjusted to compensate for the effects of wind.

If the wind is blowing, add the number of clicks circled under the flags to the rear sight windage knob setting from TRUE ZERO.

Once the windage setting is determined, record it in the WIND column.

3. Recording Data During Firing

Recording Data During Slow Fire Stages

The method for calling and plotting slow fire shots in the data book is called "the shot behind method." It allows the Marine to spend less time recording data and more time firing on the target. This is because all the calling and plotting is done while the target is in the pits being marked. This information is recorded in the DURING FIRING portion of the data book page. The proper and most efficient method for recording data during KD slow fire stages is as follows (see fig. A-2):

Fire the First Shot

Fire the first shot. Then immediately check the wind flag to see if the speed or direction of the wind changed. If the wind direction changed, indicate the windage adjustment needed to compensate for it in the WIND row under CALL number 1. This windage adjustment will have to be applied to the rifle prior to firing the second shot.

Call the Shot Accurately

As soon as the shot is fired and the target is pulled into the pits, record the exact location where the tip of the front sight was on the target at the exact instant the shot was fired (assuming sight alignment was maintained). Plot this location on the target provided under number 1 in the block marked CALL.

Prepare to Fire the Second Shot

As soon as you have recorded the call for the first shot, prepare to fire the second shot.

Look at Where the First Shot Hit

As the target reappears out of the pits, look where the first shot hit the target. Remember this location so it can be plotted after firing the second shot.

Fire the Second Shot

Fire the second shot. Then check the wind flag to see if the wind changed speed or direction.

Call the Second Shot and Plot the First Shot

As soon as the second shot is fired and the target is pulled into the pits, record the call of the second shot.

Now plot the precise location of the first shot by writing the numeral 1 on the large target diagram provided in the block marked PLOT.

Prepare to Fire the Third Shot

Repeat steps until all shots have been fired. Indicate each slow fire shot with the appropriate number (e.g., 1, 2, 3, 4, 5).

Make a Sight Adjustment if Required

Sight adjustments should be made off of a shot group, not a single shot. Determine if a sight adjustment is necessary off of the first three shots fired. If the shots form a group (i.e, a group that fits inside the "A" target bull's-eye or the center scoring ring of the "D" target), but are not where they were called, make the necessary sight adjustment.

CAUTION

Generally, major sight adjustments from established sight settings are caused by poor application of the fundamentals, inconsistencies in firing positions, inconsistencies in sight picture at different ranges and different positions, and inconsistent tension on the sling. Every effort should be made to correct shooting errors prior to making a sight adjustment on the rifle.

Note

The plotting targets in the data book for the KD Course of Fire are in inches, requiring you to calculate the number of clicks to center your shot groups.

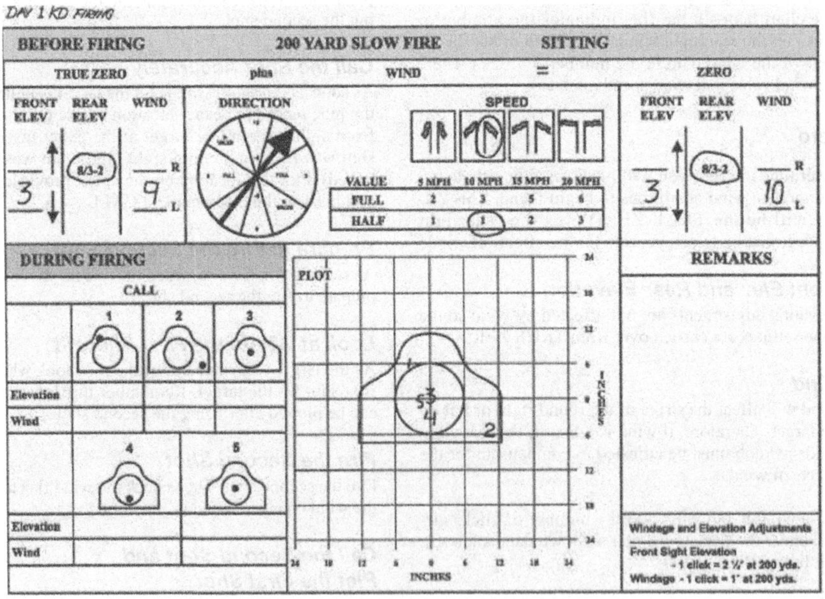

Figure A-2. Recording Slow Fire Data During Firing.

Elevation. Locate the closest horizontal grid line to the center of the plotted shot group. Follow the line across to the numbered vertical scale to determine the number of inches of elevation the shot group is off of target center. Calculate the number of clicks on your front sight post (for 200 and 300 yards) or rear sight elevation knob (for 500 yards) to bring your shot group center.

Windage. Locate the closest vertical grid line to the center of the plotted shot group. Follow the line down to the numbered horizontal scale to determine the number of inches of windage the shot group is off of target center. Calculate the number of clicks on your rear sight windage knob to bring shot group center.

Recording Data During Rapid Fire Stages

In the DURING FIRING section of the data book, record the following (see fig. A-3):

Mentally Call Shots While Firing

While firing the rapid fire string, make a mental note of any shots called out of the group.

Plot the Shot Group

After firing the rapid fire string, and when the target is marked, plot all visible hits with a dot precisely where they appear on the large target diagram in the block marked PLOT.

Make a Sight Adjustment if Required

Locate the center of the shot group. If the shots form a group, make the necessary sight adjustments off of the center of the group. If shots do not form a group and do not just contain a poor shot, do not make a sight adjustment. Determine the sight adjustment by locating the center of the shot group and using the grid lines on the target in the data book. These grid lines represent the number of inches to bring a shot group center. Looking at the shot group:

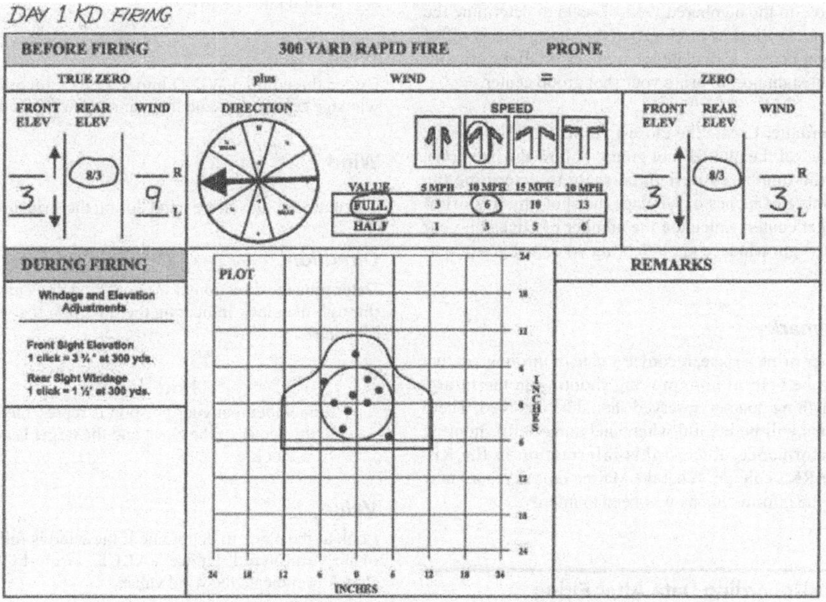

Figure A-3. Recording Rapid Fire Data During Firing.

CAUTION

Generally, major sight adjustments from established sight settings are caused by poor application of the fundamentals, inconsistencies in firing positions, inconsistencies in sight picture at different ranges and different positions, and inconsistent tension on the sling. Every effort should be made to correct shooting errors prior to making a sight adjustment on the rifle.

Note

The plotting targets in the data book for the KD Course of Fire are in inches, requiring you to calculate the number of clicks to center your shot groups.

Elevation. Locate the closest horizontal grid line to the center of the plotted shot group. Follow the line across to the numbered vertical scale to determine the number of inches of elevation the shot group is off of target center. Calculate the number of clicks on your front sight post to bring your shot group center.

Windage. Locate the closest vertical grid line to the center of the plotted shot group. Follow the line down to the numbered horizontal scale to determine the number of inches of windage the shot group is off of target center. Calculate the number of clicks on your rear sight windage knob to bring your shot group center.

Remarks

After firing a stage, record any data or information that can be helpful in improving shooting in the future. Anything done or observed should be recorded. These items will be helpful when analyzing daily shooting performance. Record this information in the REMARKS column. What the Marine fails to record may be the information he will need to improve.

4. Recording Data After Firing

In the AFTER FIRING section of the data book, record the following (see fig. A-4 on page A-6):

Zero

Upon completion of firing, determine the elevation and windage required to center the shot group (if necessary) and record this sight setting in the ZERO block of the AFTER FIRING section. If no adjustments are needed to center the shot group, record the sight settings currently on the rifle.

Front Elevation

Under the column FRONT ELEV, record the final elevation setting made on the front sight post.

Rear Elevation

Under the column REAR ELEV, record the rear sight elevation knob setting for the yard line firing from.

Note

If firing from 500 yards, record the final elevation setting made on the rear sight elevation knob.

Wind

Under the column WIND under ZERO, record final windage setting made on the rear sight windage knob.

Wind

Calculate the prevailing wind during the string of fire.

Direction

Determine the direction of the wind and draw an arrow through the clock indicating the direction the wind is blowing.

Note

Remember that your position is represented in the center of the clock and the target is at 12 o'clock.

Value

Look at the clock to determine if the wind is full, half, or no value wind. Under VALUE, circle FULL or HALF to indicate the wind value.

Speed

Observe the flag on the range and circle the appropriate flag indicating the wind's velocity (SPEED).

Determine Windage Adjustment

The chart beneath the flag indicates the number of clicks on the rear sight windage knob to offset the effects of the wind. Circle the number of clicks where the wind value and wind speed intersect.

True Zero

A true zero is the established zero without the windage adjustments to compensate for the effects of the string of fire's wind. A true zero is calculated because, the next time you fire, the wind conditions will probably be different. Therefore, the rear sight windage knob adjustments made to compensate for the string of fire's wind will not be the correct setting for wind conditions on other strings of fire or other days of firing.

Front Elevation and Rear Elevation

Because elevation adjustments are not affected by wind, the same settings are carried over from ZERO.

Wind

Calculate the windage adjustment to compensate for the string of fire's wind conditions the same way it was calculated in the BEFORE FIRING information of the data book. The only exception is now windage adjustments are being removed from the rifle rather than added to the rifle. Because the windage setting is being removed from the rifle, remove the number of clicks of windage right or left from the ZERO windage setting. Once the windage setting is determined, record it in the WIND column.

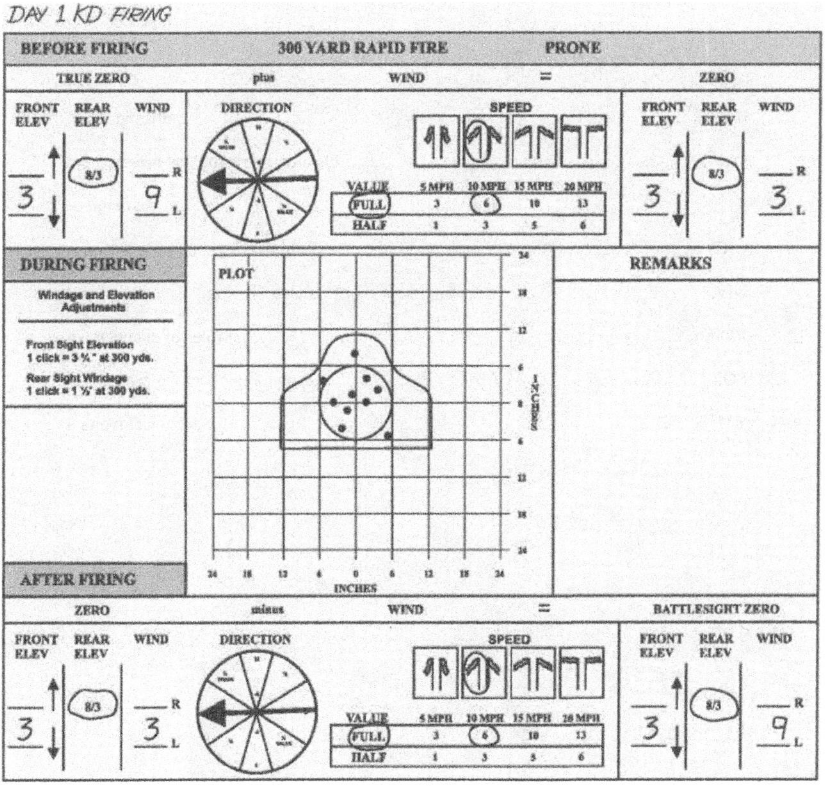

Figure A-4. Recording Data After Firing.

Appendix B

Glossary

Section I. Acronyms

BZO	battlesight zero
CLP	cleaner, lubricant, and preservative
KD	known distance
LAW	lubricating oil, arctic weapons
NBC	nuclear, biological, and chemical
ROE	rules of engagement
SOP	standing operating procedures
TM	technical manual

Section II. Definitions

A

aiming point – The aiming point is the precise point where the tip of the front sight post is placed in relationship to the target.

alibi – Any condition caused by the weapon, ammunition, or range operation that causes the shooter not to have an equal opportunity to complete a string of fire as all other shooters on the range.

B

battlesight zero (BZO) – The elevation and windage settings required to engage point targets from 0-300 yards/meters under ideal weather conditions (i.e., no wind).

bone support – The body's skeletal structure supporting the rifle's weight.

breath control – Procedure used to fire the rifle at the moment of least movement in the body and the rifle.

C

canting – An angular deviation of the weapon to the left or right from a vertical position during firing.

center of mass – A point that is horizontally and vertically centered on the target.

center of mass hold – The placement of the tip of the front sight on the target center of mass prior to the shot breaking.

centerline of the bore – An imaginary straight line beginning at the chamber end of the barrel and proceeding out of the muzzle.

chamber check – Procedure used to determine a weapon's condition.

D

double feed – Attempted simultaneous feeding of two or more rounds from the magazine.

dry fire – Cocking, aiming, and squeezing the trigger of an unloaded rifle in order to practice the fundamentals of marksmanship.

detailed search – Method for conducting a systematic search of an area for specific target indicators.

E

eye relief – The distance between from the rear sight aperture to the aiming eye.

F

function check – Procedure used to ensure the selector lever operates properly.

flag method – Procedure used to determine wind velocity and direction on a Known Distance (KD) range.

field expedient battlesight zeroing – Process used to zero the rifle at 36 yards or 30 meters when a 300-yard/-meter range is not available.

G

gas operated – A self-loading firearm that utilizes the expanding force of the propellant's gases to operate the action of the weapon.

H

hasty search – Method for quickly searching an area for enemy activity.

hasty sight setting – A zero established for distances out to 800 yards/meters.

I

immediate threat target – A target engaged at a short range (50 meters or less) with little or no warning.

initial sight setting – Sight setting placed on a rifle that serves as the starting point from which all sight adjustments are made for the initial zeroing process.

L

limited technical inspection (LTI) – An inspection performed by an armorer on a weapon to determine its operational status (safety and function, not accuracy).

line of sight – An imaginary line extending from the shooter's eye through the rifle's sights and onto an aiming point on a target.

load – Command/Procedure used to take a weapon from Condition 4 to Condition 3 by inserting a magazine with rounds.

M

magazine – A container that holds ammunition in a position to be chambered.

magazine fed – A mechanical, automatic means of supplying a firearm with ammunition to be chambered.

make ready – Command/Procedure used to take a weapon from Condition 3 to Condition 1 by chambering a round.

muscular relaxation – The state of tension required to properly control the rifle. The shooter's muscles are in a relaxed state of control—tightened but not tensed.

N

natural point of aim – The location at which the rifle's sights settle if bone support and muscular relaxation are achieved.

O

observation method – Procedure used to determine wind velocity and direction in a tactical situation.

P

pie technique – A technique used to locate and engage targets from behind cover, while minimizing the Marines exposure to enemy fire.

R

range – **1.** A designated place where live fire weapons training is conducted. **2.** The horizontal distance to which a projectile can be propelled to a specified target. **3.** The horizontal distance between a weapon and target.

recoil management – The ability to manage or control the recoil of the rifle for a given shooting position.

rollout – A technique used to locate and engage targets from behind cover, while minimizing the Marines exposure to enemy fire.

S

semiautomatic – One full cycle of operation. The firing, extraction, and ejection of the spent cartridge, the cocking of the weapon and chambering of the succeeding round of ammunition after the trigger is pulled.

shooter error — Any action induced by the shooter that causes the weapon to fail to operate properly or miss the intended target.

sight alignment – The placement of the tip of the front sight post in the center of the rear sight aperture.

sight picture — The placement of the tip of the front sight post in the center of the target while maintaining sight alignment.

sling – When properly attached to the rifle, the sling provides maximum stability for the weapon and helps reduce the effects of the rifle's recoil. Also used for individual weapon transport.

stability of hold – The ability to acquire a stable position and to hold the rifle steady for any given rifle position.

stock weld – The firm, consistent contact of the cheek with the weapon's buttstock.

stoppage – Any condition that causes the rifle to fail to fire.

T

target indicators – Anything that reveals the enemy's position.

trajectory – The path of a projectile through the air and to a target.

triangulation process – Process used to determine the vertical and horizontal sight adjustments that must be made to center a shot group.

trigger control – The skillful manipulation of the trigger that causes the rifle to fire without disturbing sight alignment or sight picture.

U

unload – Command/Procedure used to take a weapon from any condition to Condition 4.

unload and show clear – Command/Procedure used to take a weapon from any condition to Condition 4 while requiring a second individual to check the weapon to verify that no ammunition is present.

user serviceability inspection – Procedure used to ensure a weapon is in an acceptable operating condition, conducted by the shooter.

V

velocity – Speed at which the projectile travels.

W

weapons carry – Procedure used to effectively handle the rifle while remaining alert to possible threat levels.

weapons condition – Describes a weapon's readiness for live fire.

weapons transport – Procedure used to carry the rifle for long periods of time and when one or both hands are needed for other work.

windage and elevation rules – Rules that define how far the strike of the round will move on the target for each click of front and rear sight elevation or rear sight windage for each 100 yards/meters of range to the target.

Z

zeroing – The process used to adjust the rifle sights that cause it to shoot to point of aim at a desired range.